DATE DUE

MAY 0 6 2005			

Demco

LEAPS IN THE DARK

John Waller

OXFORD
UNIVERSITY PRESS

OXFORD
UNIVERSITY PRESS

Great Clarendon Street, Oxford OX2 6DP

Oxford University Press is a department of the University of Oxford.
It furthers the University's objective of excellence in research, scholarship,
and education by publishing worldwide in

Oxford New York

Auckland Bangkok Buenos Aires Cape Town Chennai
Dar es Salaam Delhi Hong Kong Istanbul Karachi Kolkata
Kuala Lumpur Madrid Melbourne Mexico City Mumbai Nairobi
São Paulo Shanghai Taipei Tokyo Toronto

Oxford is a registered trade mark of Oxford University Press
in the UK and in certain other countries

Published in the United States
by Oxford University Press Inc., New York

British Library Cataloguing in Publication Data

Data available

Library of Congress Cataloging in Publication Data

Data available

ISBN 0–19–280484–7

10 9 8 7 6 5 4 3 2 1

Typeset by RefineCatch Limited, Bungay, Suffolk
Printed in Great Britain by
Biddles Ltd, www.Biddles.co.uk

To Esther Elizabeth

CONTENTS

List of illustrations   ix

Preface   xi

Acknowledgements   xii

Introduction: The past really is another country   1

Part One: Falling from grace   **7**

  1 Joseph Glanvill: scientific witch-finder   15

  2 The man who made underpants for frogs   39

  3 Pettenkofer's poisoned chalice   63

Part Two: Eureka! revisited   **83**

  4 Sir Isaac Newton and the meaning of light   91

  5 Dr James Lind and the Navy's scourge   113

  6 The destruction of Ignaz Semmelweis   135

Part Three: Heroes made to measure   **163**

  7 Will the real Johann Weyer please stand up?   169

  8 Philippe Pinel: the reforging of a chain-breaker 191

Part Four: Do-it-yourself heroes   **213**

  9 The first casualty of war   219

  10 Rank hath its privileges   243

Conclusion: The bigger picture   267

Further reading   277

Index   283

LIST OF ILLUSTRATIONS

Frog's leap. *(James Balog/Stone/Getty Images.)*   iii

Line engraving of Joseph Glanvill by William
Faithorne after an unknown artist. *(National Portrait
Gallery, London.)*   14

The frontispiece of Joseph Glanvill's *Saducismus
Triumphatus*. Wood engraving by William Faithorne.   19

Portrait of Lazzaro Spallanzani. *(The Photolibrary.)*   38

Max von Pettenkofer. *(Science Photo Library.)*   62

Robert Koch. *(Hulton Archive/Getty Images.)*   62

Study for a 1725 portrait of Isaac Newton. *(The
Wellcome Library, London.)*   90

An idealized image of the *experimentum crucis* from the
third edition of Isaac Newton's *Opticks*, published in
1721. *(The Wellcome Library, London.)*   94

A person identified as Sir Isaac Newton showing an
optical experiment to an audience in his laboratory.
Wood engraving by Martin after C. Laverie. *(The
Wellcome Library, London.)*   108

James Lind. *(The Wellcome Library, London.)*   112

Ignaz Semmelweis. *(The Wellcome Library, London.)*   134

The Semmelweis memorial in Budapest (statue by
Alajos Stróbl), unveiled 30 September 1906. *(The
Wellcome Library, London.)*   160

Portrait of Johann Weyer, a woodcut of 1577. *(The
Wellcome Library, London.)*   168

The frontispiece of Johann Weyer's *De Praestigiis
Daemonum*, published in Basle in 1568. *(The Wellcome
Library, London.)*   171

Portrait of Philippe Pinel, a lithograph by Antoine
Maurin. *(The Wellcome Library, London.)*   190

Pinel fait enlever les feus aux aliénés de Bicêtre ('Pinel
orders the removal of iron shackles from the insane

men at Bicêtre Hospice') by Charles Müller. *(Académie Nationale de Médecine, Paris.)*   194

Robert Watson-Watt *(Bettmann/Corbis.)*   218

Selman Waksman *(Special Collections and University Archives, Rutgers University Archives)*   242

In 2002, Oxford University Press published my book *Fabulous Science: Fact and Fiction in the History of Scientific Discovery*. Its chief aim was to make cutting-edge research undertaken by historians of science much more widely accessible. The central theme of the book was that science, like other highly regarded spheres of human activity, has a tendency to create foundation myths. On close inspection, much of the received history of science, that vivid tapestry of Eureka moments, farseeing geniuses, jealous rivals, and bigoted clerics, turns out to have little basis in historical fact. Some of the most famous stories of scientific discovery are not much more than tribal tales, packed with inspirational value but extremely unreliable as accounts of the past.

As in this book, in *Fabulous Science* I employed a series of separate case studies to contrast fact with myth. Covering a wide range of fields, from physics to medicine, from Darwinism to management theory, and from epidemiology to genetics, it also sought to make a more general point: that our understanding of the history of science has been skewed by something called presentism, our tendency for judging past ideas solely by how closely they approximate to current orthodoxies. In this follow-up volume I've given greater prominence to case studies showing that those scientists who were eventually proven wrong were rarely the dullards or envious mediocrities of the standard histories. History's also-rans, often pressed into the role of fool or bigot, were often competent scientists, and their theories were perfectly good efforts for their times.

The following case studies try to cut through the layers of myth to tell more nuanced stories about the scientific enterprise. They also explore why certain individuals have achieved heroic status and how certain myths have proven so incredibly long-lasting. Once again, however, the book's main thrust is that the history of scientific discovery needs to be contextualized if we're to have any chance of understanding how we came to know so much about the world around us.

John Waller
Melbourne 2003

ACKNOWLEDGEMENTS

This book owes its existence to the labours of dozens of historians of science and medicine on whose work I have drawn. To them I owe a considerable debt. Accordingly, at the end of the book readers will find a list of the texts that inspired and informed each of the following case studies. I am also indebted to a number of friends, colleagues, and family members for their encouragement, incisive criticism, and for helping me to track down and tell stories that I feel deserve to be brought to the attention of a wider audience. I am especially grateful to Janet McCalman, Neil Thomason, Michael Waller, Richard Graham-Yooll, Richard Bellon, Matthew Klugman, and Adrian Smith.

Thanks also go to the staff of the Wellcome Trust Centre for the History of Medicine at UCL and of the Centre for the Study of Health and Society and the Department of History and Philosophy of Science at Melbourne University for creating such stimulating and congenial working environments. In addition, I have benefited from the translating skills of Alison Stibbe, the support and constructive advice of Marsha Filion, Michael Rodgers, and Abbie Headon of Oxford University Press, the comments of several anonymous referees, communications with Keith Hutchison, and the editorial work of Sandra Raphael.

The Wellcome Library in London, the National Portrait Gallery, Rutgers University, and the Académie Nationale de Médecine in Paris have kindly allowed me to reproduce several images. Ann Brothers, of the Medical History Museum of the University of Melbourne, has been of tremendous help in finding suitable illustrations. But to my wife, Abigail, I owe most of all.

THE PAST REALLY IS ANOTHER COUNTRY

Life is lived forwards, but understood backwards.

Attributed to Søren Kierkegaard

This collection of ten essays has a consistent theme: that in the history of science nothing is straightforward. Few cutting-edge experiments are ever actually decisive. Only occasionally do truths about nature reveal themselves in sudden moments of thrilling clarity. One cannot even identify a single approach that's guaranteed to keep scientists from falling into error. And although we often talk of the 'scientific method' as a unitary thing, a set of hard and fast rules that ensures progress so long as we're looking in the right direction, it really isn't like this. The early phase of developing a scientific theory entails pursuing hunches and preconceptions even in the face of unpromising experimental data. If the idea turns out to be credible then the discoverer is fêted for his or her profound intuition; if it's rejected they're likely to be condemned for overenthusiasm.

As this suggests, we're usually not terribly fair to those later judged to have got it wrong. We tend to forget how hard it is to prove a new theory and so, if they're remembered at all, we condemn the also-rans with unflinching severity. By the same token, we're often too misty-eyed about those who came up with ideas approximating to what we now believe. We overlook the fact that our heroes didn't always arrive at their ideas by a route we'd consider either rational or direct. Moreover, in many cases their evidence was anything but decisive, and their rivals absolutely correct to carry on the fight even years after what we now consider to be the truth had actually emerged. In other cases, those celebrated today as 'father figures' of one or other discipline didn't actually say what we're taught they did. Indeed, some heroic reputations are almost entirely the product of myth-makers. Occasionally, they're the result of self-aggrandizement.

Science may be the only reliable approach for making sense of the world we live in. But as these few lines suggest (and most scientists freely acknowledge), the process of discovery is a convoluted and long-drawn-out affair. Behind the trusty formula of the far-seeing hero and the story of triumph over adversity there is usually a much subtler but no less gripping reality of flawed geniuses, inadequate experiments, stabs in the dark, blind alleys, and, above all, a natural world of such treacherous complexity that brilliant men and women have been at least as likely to get it wrong as to get it right.

Unfortunately, however, the rich ambiguities of scientific research are often obscured by the way the history of scientific discovery gets told. In particular, there's a strong predilection among science writers for squeezing episodes of scientific discovery into a predictable but not very realistic three-act drama. The cast, theme, and time-scale vary, but the plot could hardly be more familiar: the hero arrives at a new idea (Act I), suffers the wrath of jealousy, conservatism, and clerical bigotry (Act II), and is then triumphantly vindicated (Act III). It's undoubtedly uplifting, but in most cases sadly lacking in accuracy.

Needless to say, not all popular accounts of scientific discovery ignore the complexities involved in doing science. Some popular science writers (one thinks particularly of the late Stephen Jay Gould) have explored with great eloquence the subtleties of the great scientific endeavour. But consider the spate of recent books with subtitles along the lines of 'The Invention that Changed the World', 'The Man who Invented Time', 'How One Man Changed the Way We See the World', and 'The Story of One Man's War Against the Establishment, Ignorance and the Deadly Scurvy'. Each is the story of, to quote from a best-selling example, 'just one man, doing it all by himself, imagining the unimaginable' and facing down 'purblind churchly certainty' and the ignorant criticism of 'dreamily unscientific' opponents. For all the many merits of these books in other respects, in each case a complex story has been shoe-horned into a single, ready-made schema. The result is that the messiness of reality is written out of existence and the ambiguities that make science so interesting are left behind.

This fondness for Great Man histories of science is regrettable because heroic dramas tell us little about how science progresses or about what scientists are up against in trying to advance our knowledge of the natural world. The aim of this book is to show, from several different angles, how misleading the notion of the lone embattled genius can be. The first three chapters look at three of history's villains or also-rans, men who advanced ideas often characterized today as irrational, pre-scientific, or just downright silly. In each case I aim to show that, given the limitations of knowledge at the time, all three formulated perfectly rational and persuasive theories. With hindsight we can see that they were misguided, but it was not through any deficiencies in reason or investigative flair that they drifted into error. Those deemed right in the long run, I will argue, have rarely had a monopoly over reason and good sense.

The next three chapters explore that staple of romantic histories, the world-changing experiment or demonstration. Examining celebrated moments in the lives of Isaac Newton, James Lind, and Ignaz Semmelweis, I attempt to demonstrate that no matter how obvious an idea might seem to us, those who come up with new theories rarely have the ability to prove them correct straight away. In most cases, it requires the development of new approaches or the refinement of old concepts, tools, and apparatus before a hunch supported by partial data can be turned into a well-attested scientific theory. Sometimes, bold new claims can only be substantiated after the originator's death. This is usually the basis for making them into far-seeing heroes spurned by inferior contemporaries. But what it really tells us, in most instances, is that our heroes weren't always sure-footed and their opponents were very often right to stick with what they'd been taught, at least until more compelling evidence was presented.

The final chapters shift the focus to the construction of heroic scientific reputations. All four subjects are revered as father figures of their disciplines, credited with opening up whole new vistas of knowledge in the fields of psychiatry, medicine, and physics. On closer inspection, however, it becomes clear that two of our quartet simply didn't say or do what most textbook accounts assert. The remaining pair *were* involved in important discoveries or developments, but they're singled out today mostly because they embellished their roles, attaching to their own names contributions actually made by others.

The main contention of this book is that we need to be very suspicious of any historical account of scientific discovery involving the romantic cliché of the man or woman before their rightful time. It's not that there haven't been individuals of exceptional ability or cases of great minds being frustrated by ignorant and mean-spirited opposition. The problem lies in our tendency to romanticize the past. There is now hardly a significant idea in the realm of science that isn't associated with a Eureka moment or a chance observation leading to an earth-shattering realization. A few of these stories are genuine, yet most are flawed to varying degrees, and many have no basis at all.

There are many reasons why the myths looked at in this book arose and were

3

then successfully propagated. Among the most important is what historians call 'presentism', a major pitfall in writing about the past, defined by Stephen Jay Gould as the 'mistaken use of present criteria to judge a distant and different past'. In history of science scholarship this trait has produced some pretty major distortions, and is in one way or another relevant to each of the following stories. Most obviously, presentism leads historians to focus only upon scientific ideas that have come to inform modern scientific paradigms. Historians must, of course, seek to explain how the present came to be as it is. But rather than trying to reconstruct how our forebears saw the world, presentist historians, professional or amateur, plunder past world-views for anything that sounds even vaguely like something we believe in today, and cast everything else aside as unimportant.

Studying the past in this way is rather like trying to write an accurate account of our planet's evolutionary past by looking only at lineages with descendants alive today. We can't hope to gain a true appreciation of the scientific enterprise unless we investigate those ideas now thought to be wrong, as well as the tiny fraction now judged to have been correct. It would be unreasonable to expect scholars to give equal time to the losers but I think it's a mistake to assume that only scientists on the winning side are worthy of our attention.

One of the reasons why some writers simplify the process of scientific discovery is that, with the benefit of hindsight, current orthodoxies and the quickest ways of getting to them can seem so obvious that we're led to assume that this must have been the route taken. Thus blind alleys, false scents, dead ends, and the circuitous and circular routes along which scientists actually travelled before arriving at their ideas about nature are overlooked. It's striking that this way of thinking about the path to discovery is even embedded in the genre of the scientific article. The Nobel Prize-winning biologist Sir Peter Medawar argued that all scientific papers are in a sense fraudulent because they deliberately smooth out the route to the conclusions they present. '[T]he scientific paper in its orthodox form,' he observed, 'does embody a totally mistaken conception, even a travesty, of the nature of scientific thought.' As Medawar himself appreciated, the path of scientific progress is rarely 'broad and straight'. By failing to acknowledge either wrong turns or the meandering routes taken, we end up with a crudely triumphalist account that does scant justice to the complexities of scientific advance.

Many of the errors associated with presentism are embodied in the way that the expression 'scientific discovery' is normally used. In most usages, the term elides three quite different things: the birth of a new idea (we might call this its *genesis*);

the accumulation of data suggesting that the idea has merit, often referred to as the stage of *verification*; and the juncture at which fair-minded peers recognize that the case has indeed been proven, or the point of *acceptance*. In old-fashioned histories, the first two steps, genesis and verification, are usually telescoped into a single instant of rationality when the clouds parted and the truth was revealed. We've already touched on some of the problems with telling the history of science in this manner. In reality, the way an important new idea is put forward may be anything but rational. Some of the most powerful concepts and organizing principles of modern science originated from obscure sources and were backed up with arguments that would today be considered absurd. As this implies, just because someone comes up with an idea that's later vindicated isn't to say that everyone ought to have downed tools and come running.

Nicolaus Copernicus' 1534 claim that the Earth revolves around the Sun provides an excellent example. We now know that in most important respects Copernicus was right. But even if his suggestion did lead to a simpler way of plotting the movements of the celestial spheres, it's hardly surprising that most of his contemporaries preferred to go on as before. Because Copernicus had no idea that the planets move in ellipses, not perfect circles, his version of heliocentrism failed to fit with the observed movements of the planets. In contrast, the long-established Aristotelian system seemed to explain every single movement of every single thing in the Universe. Copernicans were expecting people to abandon a very plausible theory of everything in exchange for an imperfect theory of just one thing. Not surprisingly, there weren't many takers.

Presentism masks these ambiguities. Having scoured history for precursors, the historian will often cast as a far-seeing hero anyone who described the world in a way even broadly in line with current knowledge. They do not always stop to consider how good their hero's evidence was at the time or how he came up with his 'great' idea. Wrongly assuming that scientific 'truth' can arise only from the most rational enquiry and that alternative theories were obviously based on sloppy research and loose conjecture, presentist historians conflate the emergence of the new concept with its becoming sufficiently credible to be worthy of wider notice. This is what gives rise to many of those canonical narratives in which a new scientific theory is said to have been devised and confirmed almost simultaneously, but then had to endure years of neglect due to the ignoble hostility of lesser minds. As several chapters in this book suggest, it's rather more common for the real hiatus to come not between the stages of verification and acceptance but between those of genesis and verification. Doubtless, many critics of new ideas have mixed motives. But the evidence so far suggests that a credible scientific theory will usually gain acceptance once a decent empirical case has been made.

Aside from their being inaccurate and unhelpful, to my mind there are three main problems with the preponderance of presentist myths in the history of science. First, in order to make a few individuals seem like gods among men, too many of history's scientists, and for that matter clerics, have been unjustly demonized. The first three chapters of this book are at least in part intended as a defence of the also-rans. Second, by asserting that most radical scientific ideas have to weather storms of ignorant and bigoted opposition, we undervalue the scientific community's impulse to criticize, correct, and dampen overenthusiasm. This is what makes it such a successful enterprise. Last, and I think most important, by disguising the true difficulties of doing science, our myths often prevent us from realizing just how impressive it is that scientists have managed to accomplish what they have. In this sense, demythologizing the history of science can only enhance our respect for scientists, both those of the present and those in the past.

PART ONE

FALLING FROM GRACE

FALLING FROM GRACE

There is no absolute knowledge. And those who claim it,
whether they are scientists or dogmatists, open the door
to tragedy. All information is imperfect. We have to treat
it with humility.

> Jacob Bronowski, *The Ascent of Man* (1973).

Our first three case studies focus on scientists from very different periods with
contrasting areas of interest: first, Joseph Glanvill (1636–80), a Royal Society
devotee who spent twenty years trying to prove that witches exist; second,
Lazzaro Spallanzani (1729–99), a brilliant Italian experimentalist who advocated
ovism, the notion that all women are born with ovaries containing millions of
fully formed, miniature human beings; and, third, Max von Pettenkofer (1818–
1901) who swallowed a flask of cholera germs because he adamantly refused to
accept the contagiousness of epidemic disease. All three stories have been
portrayed as instances in which superstition, ignorance, or plain narrow-
mindedness obstructed the path of scientific progress. Pettenkofer is now
presented as a classic scientific villain, Spallanzani as a strange paradox of genius
and foolishness, and Glanvill as a curious aberration. But on closer inspection the
careers of these men tell a rather different story, one that highlights many of the
difficulties involved both in doing scientific research and in writing a balanced
history.

After their deaths, or shortly before, Glanvill, Spallanzani, and Pettenkofer
were shown to be mistaken. But if we judge them on the basis of what it was
possible for them to know, we can easily see that all three approached the study of
nature sensibly, with due caution, and constructed entirely plausible theories.
Although their minds were constrained by prior assumptions and their objectivity
was compromised by an emotional commitment to proving their hunches right, they
were no more prejudiced or dispassionate than other scientists whose contributions
we now admire. It is indicative of this that, while in their prime, Glanvill, Spallanzani,

and Pettenkofer were highly respected by their peers and recognized as being in the forefront of their respective fields of scientific endeavour.

So what went wrong? For backing theories that subsequently fell into disfavour all three tend to be characterized as third-rate scientists. The typical attitude has been that they were wrong, ergo they must have committed some kind of logical or observational error, perhaps compounded by superstition or bloody-mindedness. The following case studies, however, do not show that the errors made by Glanvill, Spallanzani, and Pettenkofer arose from their personal failings. On the contrary, all three erred largely because of the extreme difficulties involved in studying the natural world. We *Homo sapiens* have a strong predilection for explaining things rather than merely describing them. We're continually grasping for causal theories that account for natural and social phenomena. From this comes our exceptional capacity as a species to predict, harness, and exploit. Unfortunately, these evolved capacities can also lead us into error, for when we initially attempt to comprehend all but the simplest facets of nature, we lack the tools, technologies, and conceptual frameworks required to do so effectively. Yet minds configured to spot patterns among myriads of particulars, and to discern harmonies amid what seems like meaningless noise, cannot detect whether or not a problem is soluble. If a phenomenon stands in need of an explanation, the best of the available candidates, however inadequate, is likely to be seized upon. Like nature, our minds seem to abhor vacuums.

Partly for this reason, every scientific discipline's route to the present is littered with abandoned concepts and ideas in which our predecessors once had absolute confidence. Occasionally, of course, early theorists came remarkably close to positions adopted centuries later. Francis Bacon's seventeenth-century ruminations on the possibility of continental drift and Anaximander's seventh-century BC speculations about the evolution of terrestrial life from marine creatures are examples. But in most cases, early ideas were wide of the mark. Take, for instance, Aristotelian physics, phlogiston theory, the notion of a hollow earth, and a host of theories about the medicinal advantages of smoking tobacco. What cannot be overemphasized here is that the minds that gave credence to such ideas were in no way inferior to our own. They saw questions that demanded answers and did their utmost to meet the need. Rather than mocking them, the message we need to take away is that with all humans, ourselves included, a burning sense of conviction is a far from infallible guide to what's true.

The uncomfortable fact is that believing in an incorrect theory can feel identical to believing in one that goes on to win favour. We might view with irritation what now looks like gratuitous guesswork, wishful thinking, and sloppy reasoning, but natural philosophers in the past made the best use of what they had, just as scientists continue to do in the present. Hence, for millennia, the sharpest minds of Europe, Asia Minor, and the Middle East could be as absolutely convinced of the veracity of Aristotelian physics as Glanvill, Spallanzani, and Pettenkofer were in their theories and, more to the point, contemporary scientists are in the reigning orthodoxies of our own day. There's no need to take this argument to a relativist extreme: most widely accepted theories today have a high level of predictive power. But lest one think that Glanvill, Spallanzani, and Pettenkofer should have waited and turned their attention to other matters, or even that they weren't doing science because their minds lacked the necessary openness, it's salutary to keep in mind that we are still operating with the same mental hardware as our predecessors. As such, in many areas our mental imperative to explain natural phenomena may have outstripped our present abilities to reach reliable conclusions.

The root of this difficulty, of course, is the fact that nature rarely lays out her secrets for all to see. Natural events are the outcome of infinitely complex causal chains and nothing ever happens exactly the same way twice. No wonder it's so hard to find, beneath the thick fog of discrete phenomena and events, true general principles. Given the complexity of nature and the human tendency to draw premature conclusions, it follows that adhering to the principles of the scientific method, however these are defined, is in no sense a guarantee of success. As the following chapters indicate, even the most cautious and thorough researcher can veer wildly off course.

On the positive side, scientific theorizing obviously isn't an all or nothing exercise. Especially as a field becomes more advanced, the bigger picture can be built up incrementally. Moreover, there is much to be learned from error. After all, one scientist's misconception is another's warning, perhaps even the beginning of a more fruitful line of inquiry. The physicist, Wolfgang Pauli, famously disparaged a theory with the words, 'it isn't even wrong,' meaning that an idea proven invalid tells the scientist much more about nature than theories that make no testable predictions at all. It's also the case that well-constructed scientific ideas, even if misconceived, often contain at least something of value. Thus Spallanzani benefited the science of embryology by helping to sweep aside an erroneous set of ideas that could be traced back to the ancient world. And Pettenkofer made an important contribution to the field of public health by drawing attention to epidemiological

11

patterns that had hitherto been ignored. Both men's general schemas may have been wrong, but their efforts to prove them correct threw up much of value.

In the following two chapters the word 'scientist' is used sparingly. This is because the term was not coined until 1833, when the Cambridge University polymath, William Whewell, used it in a letter to Samuel Taylor Coleridge. It did not enter general use until the last quarter of the nineteenth century. Charles Darwin, for instance, never used it to describe himself. For the periods in which Glanvill and Spallanzani worked, the older term 'natural philosopher' is much more appropriate, so I have used it instead. I have also restricted my use of the term 'science' because of the connotations it carries of a professional activity, in the sense of an occupation with its own intellectual space, a formal career structure and university-based credentials. In most places, science did not become professional in this sense at least until the later nineteenth century. I have therefore used the term 'natural philosophy' in preference to 'science' for the pre-1800 period, although the latter was at least a recognizable word to those of Glanvill and Spallanzani's generation of natural philosophers.

VERA EFFIGIES REVER: VIRI MAG: IOS: GLANVIII CAR:II REGE A SACRIS & RECTORIS DE BATH IN AGRO SOMERSET QVI VEHICVLVM MVTAVIT QVARTO DIE NOVEMB: 1680.

W: Faithorne Sculp.

DA BONA: AVERTE MALA:

Joseph Glanvill (1636–1680)

> Enlightened though he was, he was a firm believer in witchcraft, and he is chiefly remembered not as an admirer of Descartes and Bacon, and a champion of the Royal Society, but as the author of *Saducismus Triumphatus*, a monument of superstition.
>
> J. B. Bury, *The Idea of Progress* (1920).

A seventeenth-century *Who's Who of English Science* would include a handful of names familiar to us today, not least Francis Bacon, Robert Boyle, Robert Hooke, and Isaac Newton. It would also contain many more individuals recognizable only to historians specializing in the period. Unsurprisingly, only a tiny minority of the natural philosophers active during the 1600s are now remembered. The majority, who failed to discover either important new phenomena or fundamental laws of nature, have long since drifted into oblivion. Even in their own lifetimes, many among this forgotten host were fairly obscure. The same cannot be said, however, of the once highly-regarded natural philosopher Joseph Glanvill.

Three centuries years ago this son of a Devon merchant stood in the same company as Boyle and Hooke. While he was an undergraduate at Oxford University he'd written an essay extolling the distinctive approach to studying the natural world being pioneered by the Royal Society. Set up at Gresham College in London in the year 1660, the Royal Society was envisaged as a 'College for the Promoting of Physico-Mathematicall Experimentall Learning'. The natural philosophers who joined it favoured a recognizably modern approach to doing science based on careful experiment rather than armchair theorizing. It was a method designed to be empirical, democratic, and down-to-earth, and Glanvill revelled in it. Published in 1661 as *The Vanity of Dogmatising*, his undergraduate essay

LEFT: *Line engraving of Joseph Glanvill by William Faithorne after an unknown artist.*

won him immediate membership of the Royal Society itself. Thereafter, when not ministering to his congregation in the provincial town of Bath or serving as chaplain to King Charles II, Glanvill acted as one of the Society's most effective publicists. He may have performed no memorable experiments himself, but he did something perhaps just as important: he campaigned long and hard to establish the experimental approach to doing natural philosophy.

The Royal Society that Glanvill promoted was at the forefront of what's now termed the 'scientific revolution'. Covering the period from the completion of Nicolas Copernicus' *De Revolutionibus* in 1543 to the death of Isaac Newton in 1727, most historians agree that this dramatic phase in the history of science saw the first flowering of modern scientific attitudes. One of the greatest accomplishments of the Scientific Revolution is said to have been the demise of the superstitious belief in witchcraft. The argument goes that witches have never been more than a figment of human imagination, but as people before the Scientific Revolution lived in a world they couldn't properly understand, they imagined the existence of all manner of supernatural forces in order to make sense of it. The likes of Galileo, Boyle, and Newton then put forth rational explanations for what had previously been inexplicable phenomena. These, at last, weaned people off the rag-bag of superstitious nonsense that had barred human intellectual progress for centuries. The very idea of witches became a grotesque absurdity and witch-burning a byword for the darker side of human nature.

Knowing that Glanvill was a standard-bearer for the Royal Society and a staunch advocate of experiment rather than lofty speculation, one naturally assumes that he battled ferociously against all forms of superstition. But this is where things become problematic, for Joseph Glanvill threw his entire weight behind the legendary *enemies* of science and reason: the witch-finders. Starting with a 1666 book entitled *A Philosophical Endeavour towards the Defence of the Being of Witches and Apparitions*, this Royal Society stalwart established himself as one of the most famous believers in witchcraft his generation produced. *A Philosophical Endeavour* was made up of two parts: the first argued that it was reasonable to suppose that men and women really do forge pacts with the Devil, and the second cited a series of strange incidents as proof that they did.

Unfortunately for Glanvill, in 1666 most copies of the book were incinerated in a printer's warehouse during the Great Fire of London. Undaunted, he developed and refined his case in several more works. All were targeted at the growing number of people who thought belief in witches fit only for children and rustics. Despite being the author of a book entitled *Scepsis Scientifica*, which celebrated the critical spirit of Royal Society science, Glanvill expressed only scorn for those who denied that witches were real. Nor were there any clear limits to the devilish abominations he accepted. Glanvill included in his books everything from witches flying through windows, killing at a glance, inducing convulsions, turning themselves into 'cats, hares and other creatures', to their raising storms at sea and suckling satanic familiars.

Glanvill seems to present us with something of a paradox. A maestro of scientific scepticism, he nevertheless succumbed to a set of beliefs already considered outdated by most educated contemporaries. Until recently, when historians discussed him, his belief in witchcraft was understandably portrayed as a kind of personal weakness and his conversion to modern rationalism pitiably incomplete. As the quotation from J. B. Bury's classic *The Idea of Progress* at the head of this chapter indicates, Glanvill was seen to have let the side down, wasting his talents on the impossible task of riding two horses heading in opposite directions. But the bigger question raised by Glanvill's curious career is why the Royal Society tolerated one of its members spending so much of his life trying to prove the existence of witches. Why did men like Robert Boyle befriend and admire a man so manifestly hooked on superstition?

In this chapter we'll see that far from being regrettable aberrations, during the late 1600s Glanvill's books on witchcraft could lay claim to absolute scientific integrity. He may have been wrong (to our minds at least), but the case he presented for the reality of witches bore all the hallmarks of Royal Society science. Furthermore, in key respects, Glanvill's reasoning was identical to that employed by such luminaries of the Royal Society as Boyle and Isaac Newton in their now celebrated analyses of air, light, matter, and planetary motion. Though it may be hard to believe now, there was nothing in the least bit superstitious or unscientific about believing that witches were real. Because we moderns no longer take notions of witchcraft seriously, we're led to surmise that no serious natural philosopher of the 1600s could have done so. But as this chapter argues, even though our 'scientific attitudes' seem to preclude a belief in witchcraft, back in the seventeenth century the existence or otherwise of witches was an open, empirical question that at least some respectable philosophers saw it as their duty to investigate, and they did so in an entirely rational manner.

The Drummer of Tedworth

In 1664 a Justice of the Peace called Robert Hunt claimed to have uncovered a coven of witches in Somerset. Having brought charges against them, Hunt was astonished to find that most of the local gentry responded with complacent disdain. Accused witches had been undergoing ducking, torture, and burning as recently as twenty years earlier, but there was fast-growing cynicism about the probity of witch trials and the reality of witchcraft. More than eighty years before Hunt's revelations, the Kentish county squire and hop-grower, Reginald Scot, had written a book heralding this change of heart. He'd rejected the possibility of supernatural events in his *The Discoverie of Witchcraft*, and poured scorn on the 'Cruelties, absurdities and impossibilities' of the self-appointed witch-hunters. 'Oh absurd credulitie!' was Scot's constant refrain.

Those who mocked Robert Hunt cited cases such as that of France's Marthe Brosier, who had claimed in the 1590s to be a witch. Under examination by physicians sent by the king, it quickly transpired that she was no more than a clever trickster. In one test the doctors splashed her with holy water, having told her that it was unconsecrated. She showed no distress. Then they applied plain water which they said had been blessed. Madame Brosier, knowing that blessed water was supposed to cause witches terrible suffering, fell writhing to the floor. She'd walked straight into the doctors' well-laid trap. 'Nothing from the devil, much counterfeit, a little from disease,' was the chief physician's trenchant analysis. And, in part due to the unmasking of dozens of similar frauds, by the 1660s the educated classes of Europe had mostly decided that God was the only worker of miracles and that he'd now sheathed his wand.

The frontispiece of Joseph Glanvill's Saducismus Triumphatus. Wood engraving by William Faithorne.

Glanvill, however, was far less sure. Appalled at the complacency of Hunt's detractors, he published *Some Philosophical Considerations touching Witches and Witchcraft*. The copies that survived London's Great Fire were so well received that in 1668 he issued a much expanded edition entitled *A Blow at Modern Sadducism* ('Sadducism' was Glanvill's term for those who denied the reality of spirits). An even larger version, *Saducismus Triumphatus*, was published in 1688, several years after his death. In researching these volumes, Glanvill had undertaken personally to investigate claims of possession and witchcraft. For example, in late January 1663 he visited the house of one Mr. Mompesson in Tedworth, Wiltshire. This was the site of an alleged haunting by a 'drumming' devil that was already sufficiently notorious to have become the subject of a popular song. The story went that in March 1661 Mr Mompesson had had the drum of a busker confiscated in the neighbouring town of Ludgarshal after its owner had spent several days illegally harassing local people for money. When the seized drum was brought to Mr Mompesson's house it seemed to trigger a wave of supernatural activity. Strange noises such as the loud drumming of well-known tunes, like *Round-heads and Cuckolds* and the *Tat-too*, began to be heard late at night and in the early hours of the morning. In the following weeks, the hauntings seemed to escalate.

Beds shook violently, boards were seen floating in the air, and chairs moved for no apparent reason; sounds were heard of panting dogs, purring cats, and clinking money; chamber-pots were emptied on the children's heads; a naked knife appeared in the bed of the mistress of the house; and claw impressions were left in the fire ash. Next, Bibles were found upturned and opened to pages concerning Old Testament demons. Shortly after, strange animals and ghostly figures were seen in the house and, on one terrifying night, a vague form 'with two red and glaring eyes' appeared to members of the household. By the time Glanvill arrived, the haunting had been going on for nearly two years. But the Devil, he later noted, 'had ceased from its Drumming and ruder noises before I came thither'. Nonetheless Glanvill was treated to plenty of 'strange scratching' sounds and the sight of the children's beds moving without visible cause. In addition, Glanvill's groom woke him early on the first morning after his arrival with the news that his horse was in a terrible sweat. During the course of a gentle ride that morning, it suddenly expired. Glanvill later wrote that he had made every possible effort to discover if trickery were afoot, but found not the slightest evidence of deceit. The cause of the strange events, he concluded, could only be 'some *Daemon* or *Spirit*'.

Not long after, the erstwhile drummer of Ludgarshal was apprehended for theft. While in prison he was said to have boasted to a Wiltshire man that he had 'plagued' Mr Mompesson for taking his drum. 'And he shall never be quiet,' the drummer added malignly, 'till he hath made me satisfaction for taking away my Drum.' On the strength of this reported conversation the man was tried for witchcraft at Sarum (Salisbury), found guilty, and sentenced to transportation. Glanvill was among those who gave sworn evidence against him. For some time after, peace finally returned to Mr Mompesson's house. But, by all accounts, the drummer escaped and, for a time at least, the haunting began afresh.

The 'drummer of Tedworth' was only one of many narratives of hauntings and witchcraft recounted in Glanvill's *Saducismus Triumphatus*. The next account in sequence concerned a slain woman who came back from the grave to tell of her murder and to reveal the whereabouts of her hidden corpse. In the story that followed, a young boy was struck down with illness, having been touched by a local woman, Jane Brooks, and subsequently fell into convulsions whenever she came near him. Yet another account described the then famous Irish witch of Youghall, who was said to have used her diabolical powers to subject one Mary Longdon to 'Fits and Trances' and to have later killed a man whose hand she kissed through the bars of her gaol window. In all these cases, Glanvill drew upon testimonies given by dozens of different parties at assize court sessions. Unfortunately, the fate of the Youghall witch is not described, though Jane Brooks, we are told, was executed for witchcraft on 26 March 1658. Glanvill did not comment on whether he approved of capital punishment for convicted witches, but about the guilt of both Jane Brooks and the Youghall 'Hagg' he had not the slightest doubt.

Of confidence and diffidence

To modern readers, as to some of his contemporaries, Glanvill's stories are strongly redolent of gullibility. In his *Saducismus Triumphatus*, however,

Glanvill absolutely insisted that the scientific deficiencies were all on the side of his opponents; thus he mercilessly denounced the critic who '*swaggers*, and *Huffs*, and *swears* there are no *Witches*'. This is not an easy accusation for us to take seriously. Nevertheless, if we're to avoid the pitfall of presentism we must make every effort to understand why Glanvill believed his conclusions to be more rational than those of his critics. To do this we need to take a brief look at the principles that defined Royal Society science. And, in the later 1600s, there were few more eloquent exponents of the new scientific attitudes than Glanvill himself.

In 1661 he identified two chief obstacles to scientific progress. These were 'extreme Confidence on the one hand, and Diffidence on the other'. Let's start with the sin of overconfidence. Natural philosophers were being warned to avoid having too much faith in the veracity of their theories. 'Confidence in uncertainties,' Glanvill wrote, 'is the greatest enemy.' The reason for this, he elaborated, is that we have feeble minds, imperfect senses, and strong passions: a recipe for getting things wrong and failing to recognize it. Scepticism, therefore, had to be cultivated by all serious philosophers. But Glanvill also recognized that total scepticism leads to paralysing defeatism. A certain amount of 'confidence' is also therefore essential, for if we place no trust at all in our senses then we'll give up long before trying to make sense of the world.

The solution, strongly endorsed by Glanvill, was for the natural philosopher provisionally to accept some things as correct but always to be ready to abandon them in the light of further experience. Science was to involve a balancing act between 'confidence' and 'diffidence'. Nothing was to be accepted as absolutely true, since one can never justify feeling totally sure about any single interpretation. But in certain circumstances an idea could be declared 'beyond reasonable doubt'. Everything, in short, was to come down to probability judgements—what Boyle referred to as 'probationary truths' or 'moral certainties', and John Wilkins, one of the founders of the Royal Society, described as requiring a sense of 'conditional infallibility'. But how could the natural philosopher know when it was right to advance from initial scepticism to provisional confidence?

In this respect, nearly all Royal Society regulars were in agreement: the only basis for having confidence in the veracity of an observation or theory was for a large number of competent persons to witness a phenomenon and then collectively agree on what they'd seen and how best to interpret it. In this way, it was claimed, natural philosophers would restrain their imaginations and be able to come to rational conclusions. Nobody should trust what only one person claims to have seen, but there's no such difficulty with a room full of respectable and intelligent witnesses. Hence, Thomas Sprat, the chief apologist for the Royal Society, explained that it was 'the union of eyes and hands' that gave its Fellows legitimate 'confidence' in their scientific claims. Witnessing, in effect, provided the checks and balances needed to counteract arbitrary theorizing and errors of judgement. It was the antidote to scientific closed minds and intellectual parochialism.

What we're describing here is, of course, the idealized mindset of modern scientific inquiry. As we'll see in more detail in the chapter on Newton, even if today scientists seldom gather together to witness each other's experiments, they strive for the same effect by carefully replicating them in their own laboratories. And, just as in the 1600s, an experimental phenomenon observed by many competent persons is deemed to constitute a real event.

This now seems commonsensical, but in the 1600s it was all rather innovative. This isn't only because experiments had previously been the exception rather than the rule (Galileo is unlikely to have performed the 'Leaning Tower of Pisa' experiment anywhere other than in his own mind). More important, philosophers in the past had had little time for ideas about which they couldn't already feel certain. Why, they asked, settle for anything less than absolute certitude? But if the members of the Royal Society sounded modest in denying that absolute certainty was attainable, it's important to note that they felt there were few limitations on what they'd eventually be able to explain about the natural world. Brimming with self-confidence, they declared all the great mysteries of nature—magnetism, ocean tides, reproduction, and the movement of heavenly bodies—open to rational investigation. Typical of the period's bold and often sexual rhetoric, Robert Boyle predicted that the new experimental science would allow students of nature to extract a 'confession of all that lay in [nature's] most intimate recesses'.

And, to his credit, Boyle showed how it ought to be done. A first-rate experimenter, he used the recently invented air pump to draw air out of a glass vessel. In so doing he overturned the orthodoxy of the day by demonstrating that vacuums are a physical possibility. He went on to reveal that sound does not travel in a vacuum, that fire cannot be produced in the absence of air, and that there is a close relationship between the volume and the pressure of any quantity of gas. It was an experimental tour de force. And, in each case, Boyle's ideas achieved credibility because they were carefully staged before gatherings of competent witnesses who mostly gave their assent to his interpretations. As Boyle fully appreciated, in the hallowed meeting rooms of the Royal Society a theory stood or fell according to the consensus of the assembled observers. Nothing was proven until the members had seen it for themselves.

But what do the exacting standards of Royal Society science and Boyle's experiments with the air pump have to do with Joseph Glanvill's bogus stories about witches? A clue is contained in the fact that Boyle himself sincerely supported Glanvill's investigations into witchcraft phenomena. Writing to Glanvill on 18 September 1677, he enjoined his colleague to collect together a greater abundance of 'well verified . . . testimonies and authorities' of hauntings, possessions, and devilry. Boyle went on to say that he suspected nineteen out of twenty witchcraft narratives to be fraudulent, but forcefully added that just one story based on confirmed testimony would persuade him that witches were real. Boyle's message was clear: whether witches existed or not was a straightforward empirical question to be tackled in the same manner as one would research the possibility or otherwise of vacuums. As such, the matter had to be investigated on the basis of 'sensible evidence' provided by credible witnesses. Royal Society protocols were to be applied to studying the most mysterious phenomena in nature.

'The garb of the naturalist'

That Glanvill went about investigating the subject of witchcraft as a Fellow of the Royal Society could hardly be clearer in his writings. Each of his works on witchcraft began with a defence of the validity of expert testimony in science. Few of us, Glanvill wrote, have ever been to Rome or Constantinople, but we have no doubt that these cities exist. We are willing to take a lot on trust, he went on, because we know that at least 'some Humane Testimonies are credible and certain [since] all Mankind are not Lyars, Cheats, and Knaves.' In the case of witchcraft, Glanvill knew that not everyone could be trusted to tell the truth. There were obviously numerous cases of deception; Madame Brosier was not an isolated offender. But, he observed, the fact that some alleged hauntings have been proven specious says nothing about the veracity of others. It is 'bad Logick,' Glanvill wrote, 'to conclude in matters of Fact from a single Negative . . . By the same way of reasoning,' he continued, 'I may infer that there were never any Robberies done on *Salisbury* Plain . . . because I have often Travelled all those ways, and yet was never Robbed.' The testimony of reliable men and women was no less valid just because some people in the past had perjured themselves.

The natural philosopher's duty was to go out and seek honest testimony while always being on his guard against fraud. Given that the agreement of reliable witnesses was an accepted standard of proof at the Royal Society, it stood to reason that it was sufficient in this area as well. This is why Boyle advised Glanvill to collect 'well verified' narratives and why Glanvill spent almost twenty years trying to do just that. To both men it seemed entirely obvious that if strange noises, levitating objects, and flying hags were witnessed by credible individuals, and no 'natural' explanation seemed possible, then the reality of witchcraft had to be accepted. Accordingly, Glanvill explicitly based his case on the 'Attestation of Thousands of Eye and Ear witnesses, and those not of the easily deceivable Vulgar only, but of wise and grave Discerners'. In the 'Garb of the Naturalist', to borrow the expression used by his ally, Henry More, Glanvill painstakingly collected 'the clearest testimonies [above] the conceits of our bold imaginations'.

Glanvill's account of 'Mr. Mompesson's disturbance' exemplifies the no-nonsense rationality of his approach. While telling the story of the Tedworth haunting as others might have done, Glanvill went much further than anyone else in carefully enumerating the many different persons who had seen or heard the Devil and his minions in Mr Mompesson's house. These included several members of Mompesson's family, his household servants, his neighbours, several ministers of the

Church, and a variety of visiting worthies, including a son of Sir Thomas Bennett, Sir Thomas Chamberlain of Oxford, Glanvill himself, and a personal friend named Mr Hill. Similarly in line with Royal Society protocol, he placed little confidence in anything observed by only a single person. For example, in *Saducismus Triumphatus* he wrote of having been the only one to see the apparition of a small creature in Mr Mompesson's house. He thus confessed that since it was based upon his 'single Testimony', he'd held this evidence back for a number of years as less trustworthy than collectively witnessed events. The main weight of Glanvill's case always rested on those strange occurrences observed by several people at the same time or witnessed by numerous different people at different times.

Again strictly in accordance with the procedures of the Royal Society, once everyone had agreed upon the diabolical nature of what they had seen in Tedworth Glanvill asked whether their testimonies were trustworthy and reliable. This was a crucial question because he knew there were many reasons why a person might be led to make a false statement in the matter of witchcraft. But in the case of the ghostly drummer of Tedworth, as in other alleged hauntings he studied, Glanvill was able to employ plausibility criteria and reject them one by one.

Glanvill accepted that his witness statements would be less dependable if those who gave them had gone to Tedworth already accepting the reality of witches. This would have rendered them gullible and easy prey to cunning tricksters. But, he insisted, such was not the case. He could personally vouch for the open-mindedness of some of the house guests. Mr Hill, for instance was 'a very sober, intelligent, and credible person'. And of the other witnesses, '[m]any came prejudiced against the Belief of such things in general and others revolted before-hand against the Belief.' Of course, we have to take Glanvill's word on this, but as the reaction to Robert Hunt implies, there were plenty of unbelievers in provincial England; some of those who visited Tedworth must have harboured serious doubts. And, if Glanvill is to be believed, many of these sceptics overcame their prejudices, having spent a night or two under Mr Mompesson's roof.

Then came the possibility of insanity on the part of at least some witnesses. After all, Glanvill noted, seeing ghosts is sometimes no more than an 'illusion of crasie Imagination'. So could Mr Mompesson have been hallucinating? This wasn't likely since he was known to be a level-headed kind of fellow. And even if he were subject to the kinds of 'melancholy humours' then believed capable of inducing hallucinations, this would hardly explain the unambiguous testimony provided by his family and neighbours, not to mention a series of very intelligent and in some cases highly sceptical guests.

Fraud remained the most likely explanation. Perhaps Mr Mompesson was thumbing his nose at his neighbours by gulling them into absurd beliefs. Glanvill, however, vehemently denied any possibility of chicanery on Mr Mompesson's part. Not only did he seem to be of the finest moral character ('neither vain nor credulous, but a discreet, sagacious and manly person'), but he had suffered grievously 'in his Name, in his Estate, in all Affairs, and in the general Peace of his Family' as a result of the alleged haunting by the drumming Devil. He may have gained plenty of attention as a result of it, but almost all of this attention was of the most undesirable kind: the many unbelievers took Mr Mompesson for a fraud and even most believers assumed that he'd committed a deeply impious act and had brought upon himself the wrath of God. From a more practical point of view, for several years he struggled to find servants willing to work in his spirit-troubled house. Surely this man could not be suspected of putting 'a Cheat upon the World'.

Glanvill then reflected on Mr Mompesson's servants. In this age of levellers and republican experiments, it was surely conceivable that a disgruntled employee had taken a long-drawn-out revenge against an overbearing employer. Glanvill disposed of this possibility by arguing: 'It cannot with any shew of reason be supposed that any of his Servants abused him since in all that time [Mr Mompesson] must needs have detected the deceit.' The master of the house couldn't possibly have been fooled hundreds of times without the least suspicion. And it would have taken extraordinary ingenuity for a servant to have caused: 'the Motion of Boards and Chairs', 'the beating of a Drum in the midst of a Room, and in the Air', and 'the violent Beating and Shaking of the Bedstead, of which there was no perceivable Cause of Occasion', all without being detected. Had servants been responsible, at least one of the many 'Jealous' and 'Inquisitive' persons present on the hundreds of occasions when noises were heard or movements seen would have uncovered the ruse. This left only Mompesson's family, and they were unquestionably free from sus-picion. As Glanvill noted: 'what interest could any of his Family have had . . . to continue so long so troublesome, and so injurious an Imposture?'

In short, Glanvill saw no reason to suspect deceit in Mr Mompesson's household. Dozens of reliable persons had been prepared to confirm the story to Glanvill himself *and* under oath before the assize court that had sentenced the drummer to deportation. To Glanvill these witnesses were essentially analogous to those who'd gathered around Boyle's air pump to see for themselves whether there was any truth to his startling claims about the existence of a vacuum and the properties of air. Several had even endeavoured to find evidence of deception and yet had failed utterly. It therefore seemed churlish to deny that witchcraft was the true cause of the disturbances. As Glanvill wrote, some 'invisible extraordinary agent' must have been responsible for the bizarre events, and this was the only rational explanation for the fact that 'these transactions were *near* and *are public, frequent, and of diverse years continuance*, witnessed by multitudes of competent and unbiased attestors, and acted in a searching and incredulous age: Arguments enough, one would think, to convince any modest and capable reason.' The strange phenomena had been witnessed by competent persons and those who rejected these statements, Glanvill added, are just 'as void of *reason*, as they are of *charity* and *good manners*'.

Glanvill felt able to go on the attack because he was confident that his investigations had conformed exactly to the procedures of Royal Society science. And so, having completed his first major study of witches, he felt entitled to ask the Royal Society to 'direct some of its *wary* and *luciferous* enquiries towards the *world of spirits*'. No official response was forthcoming, but Glanvill did receive plenty of personal support. Boyle became a powerful ally. In 1672 he sent Glanvill a detailed report of an alleged Irish witch whose powers he had personally verified. Boyle also confirmed the authenticity of a haunting by a poltergeist in the town of Mascon (Macon) in France and discoursed at length on the alleged phenomenon of 'second sight', the unnerving capacity some people appeared to have of being able to tell the future. Moreover, so strong a case for the existence of witches did Glanvill build that his work became famous for persuading the most hardened doubters that witches were real.

Samuel Pepys, an arch-pragmatist, was so impressed by the quantity and quality of Glanvill's evidence that he described *Saducismus Triumphatus* in his diary as 'worth reading indeed'.

Land of spirits

Glanvill was neither painfully credulous nor wandering upon the margins of legitimate scientific theorizing. He may have been fooled in Tedworth, but there was little wrong with his logic and nothing remarkable in his obtaining the unqualified support of dedicated experimentalists like Robert Boyle. But surely, one could still argue, there was something implicitly ridiculous in the idea of witchcraft. No matter how many people testified to witnessing moving beds, upturned Bibles, and fiery eyes, a sensible person would have known that they were mistaken. From this point of view, *Saducismus Triumphatus* might be seen as an unjust victory of premature confidence over proper scientific diffidence. So was Glanvill a fool for taking the idea of devilry seriously in the first place? Or, to put it in terms a modern philosopher might employ, were there strong *a priori* reasons why a seventeenth-century Fellow of the Royal Society should have recognized accounts of witchcraft, hauntings, and strange apparitions as nonsense?

In answering this question we need to remember that Joseph Glanvill was not writing in a highly secularized time, like our present. Virtually all Royal Society philosophers believed without reservation in the existence of a spirit world inhabited by God, Christ, and numerous angels, as well as the souls of all the men and women deemed pure enough to enter the kingdom of heaven. That another, purely spiritual, dimension existed was for most natural philosophers an indisputable fact. Furthermore, many of the leading figures of Royal Society science saw it as their duty to help combat the rising tide of atheism detectable in the coffee-houses and the clubs of the metropolis. Isaac Newton, for instance, had anything but conventional religious views, but when he mistakenly came to believe that the philosopher John Locke was an atheist, he expressed a sincere wish that Locke would meet an early death.

Boyle and Hooke also expressed strong religious convictions. Indeed, they even argued that Royal Society science legitimized Christian, and especially Anglican, theology. As for Glanvill, he openly admitted that the chief attraction for him of proving the reality of witchcraft was the aid it would give him in stamping out unbelief. Simply prove that witches are real, he calculated, and a belief in a spirit world would naturally follow. For, he argued, the existence of witches automatically implies that there exist 'other intelligent Beings besides those . . . clad in heavy Earth or Clay'. Indeed, religion seems to have brought Glanvill considerable comfort. As he put it in the epistolary dedication of *Saducismus Triumphatus* to Charles II: '*there is nothing can render the thoughts of this* odd life tolerable, *but the* expectation of *another.*' But he was hardly unusual in believing so ardently in the Christian God. One of the few things upon which Royal Society natural philosophers did agree was the existence of heaven. Nor did they think the divine kingdom the only realm in which the spiritual held sway.

Until as late as the 1980s, most students taking courses in the history of the Scientific Revolution were informed that natural philosophers of the period ruthlessly expunged from science any talk of immaterial forces on the basis that these smacked of old-fashioned superstition. All respectable philosophers, until Isaac Newton, were said to have denied the existence of forces acting mysteriously on bodies from a distance. Instead they strove for models which explained everything in terms of simple mechanics. But in recent years historians have demonstrated that this view is simply wrong. Nearly all natural philosophers of the time, including those at the Royal Society, felt that there had to be at least some immaterial or spiritual activity in the universe. Otherwise they could see no way of accounting for such strange phenomena as magnetism, electricity, the bonding of particles of matter, and, not least, human thought.

Boyle, for instance, wrote of the importance of 'a very agile and invisible sort of fluids, called spirits, vital and animal', which explained such things as magnetism and even life itself. However, as both Newton and Boyle made clear, there was nothing miraculous about the immaterial forces they were invoking. Neither was willing to believe in spiritual forces said to act randomly or haphazardly. The only kinds of force that natural philosophers felt able to speculate about were those that, like gravity, could be said to obey strict natural laws. Anything else was seen as superstitious. Nor, crucially, did this kind of rubric place witchcraft beyond the scientific pale. In fact, in the tradition of medieval demonology, Satan and his minions were not believed to act 'supernaturally' in the sense of being able to subvert nature. They were thought to accomplish extraordinary feats using immaterial forces, but *always* in accordance with the fixed laws of the universe. While God could transcend these laws at will, the much lowlier Devil, as befitted a fallen angel, could at best act 'preternaturally'. This meant that while his profound knowledge of the world gave him tremendous power, everything he did was constrained within the bounds of natural law.

The same went for the men and women who were in league with him. As Glanvill put it, witches go beyond 'the common *road*' of nature, but they're simply making use of the greater knowledge and mastery of nature's laws imparted to them by Satan. Devilish magic was powerful but natural. By implication, once the experimental method had extracted from nature more of her secrets, humans too would be capable of the same astonishing deeds. 'To them that come after us,' Glanvill enthused, 'It may be as ordinary to buy a pair of wings to fly to the remotest regions, as now a pair of boots to ride a journey.' The key point is that if Satan's power was said not to be miraculous, then believing in witchcraft in the late 1600s was fully consistent with the ethos of the Scientific Revolution. If Boyle's 'agile and invisible sort of fluids' could be believed in by mainstream natural philosophers, so could the equally natural forces associated with witches.

Yet Glanvill did go further than many of his peers by arguing that Satan is still active in the world. And so one response might be that he should have invested more energy in uncovering evidence of possible fraud. This, however, would be neither fair to Glanvill nor realistic about the nature of the scientific enterprise. It's widely accepted that scientists tend to have too much invested in their theories to be effective in spotting their flaws. This is one reason why science needs to be a community activity: competition among scientists acts as a powerful corrective to individuals getting carried away with their own narrow explanations. Moreover, we cannot specify a degree of scepticism that's necessary for good science. Ranging over some of the multitude of scientific discoveries made to date, it's clear that progress has sometimes required a scientist to place tremendous confidence in an idea for which the evidence was, at first, highly ambiguous. Indeed, the philosopher of science, Gerald Holton, argues that scientists working in controversial areas need to be able to 'suspend their disbelief', sticking to a theory even when the first data-sets don't look very promising. Judgements about whether a scientist was insufficiently sceptical must therefore be cautious and pragmatic.

In the case of Joseph Glanvill, I think there can little doubt that he displayed an acceptable quotient of scepticism. Glanvill was unquestionably committed to a belief in witches before he began his investigations, but given the scientific integrity of ideas of spirits and immaterial forces in the 1600s, and the apparent reliability of the men and women who swore to having seen bizarre things in Tedworth and elsewhere, his belief in witches seems no more fanciful than the conclusions reached by countless other bona fide scientists, before and since.

The force of argument

We've seen that Glanvill was convinced his opponents were committing a gross scientific error by assuming, rather than proving, that witches don't exist. Their 'mighty Confidence [is] grounded upon nothing', he stormed in *Saducismus Triumphatus*; they brush aside volumes of expert testimony without so much as a pause for reflection. Glanvill surmised that the rationale upon which these sceptics denied the reality of witches was quite simple: they could imagine no plausible means by which any of the bizarre acts attributed to witches could be accomplished. Witchcraft was unintelligible and therefore impossible. Or, in Glanvill's own words, the activities of witches were doubted only 'because there [were] difficulties in the conceiving of it'. Glanvill tackled this argument head-on and in doing so played what must be seen as his finest hand.

Nature, Glanvill explained, is full of impenetrable phenomena. For instance, he elaborated, '[w]e cannot conceive how the Foetus is form'd in the Womb, nor as much as how a Plant springs from the Earth we tread on.' Yet babies are indubitably born and seeds routinely produce plants. And despite the fact that no one can imagine how new life is created, the well-attested nature of the phenomenon itself is enough to convince people that it really happens. One can imagine Glanvill pausing dramatically at this point and then posing the key question: why should witchcraft be treated any differently? He had managed to gather together the credible testimonies of dozens of people all professing to have witnessed devilry in action. In his view at least, by the rigorous standards imposed by the Royal Society this ought to have been enough to establish witchcraft as a true phenomenon. And, since witchcraft was no more or less intelligible than the generation of new life, there was no justification for taking one for granted and yet denying the other. As Glanvill wrote in *Saducismus Triumphatus*:

> [we] cannot conceive how such things [as witchcraft] can be performed; which only argues the *weakness* and *imperfection* of our *knowledge* and *apprehensions*, not the impossibility of those performances: and we can no more from hence form an Argument against them, than against the most *ordinary effects in Nature* [like generation].

Glanvill's argument boiled down to the simple assertion that '*matters of fact* well proved ought not to be denied.' It doesn't matter if an alleged phenomenon makes sense or not; all that's required for it to be accepted as real is for it to have been confirmed by the 'immediate Sense, or the Testimony of others'. Denying the reality of witches in the face of the expert testimony he'd amassed was:

> poudly to exalt [one's] own *opinions* above the clearest *testimonies* and most sensible *demonstrations* of *fact*: and to give the *Lye* to all *Mankind*, rather than distrust the *conceits* of [one's] own bold *imaginations*.

Until contrary evidence emerged, no one was entitled to reject out of hand the statements of reliable witnesses.

Because Glanvill deployed these arguments in support of his now discredited belief in witchcraft, some readers may be inclined to dismiss them by association. It is therefore instructive to look ahead briefly to the way one of the undoubted geniuses of the period justified *his* belief in immaterial forces. When Isaac Newton first proposed the theory of universal gravitation in 1686 many natural philosophers were deeply

33

sceptical. After all, they pointed out, it's entirely unintelligible how this hypothetical force could work. How could one star or planet possibly affect another from a vast distance? Newton consistently replied to these critics that it need not concern us that gravity's 'Causes be not yet discover'd'. It only matters that the 'Truth [of gravity's existence] appear[s] to us by Phaenomena'. In other words, the movements and positions of heavenly bodies prove that gravity exists and, as such, the mysteriousness of the force is neither here nor there. 'I scruple not to propose the Principles of Motion above-mention'd,' Newton wrote in his *Opticks* of 1704, 'they being of very general Extent.' Since clearly observed and widely verified phenomena indicated that this hypothetical force was real, it made no difference at all if anyone could say how it worked.

Self-evidently, Newton's line of reasoning is indistinguishable from that used earlier by Joseph Glanvill to prove the existence of witches; it may even be that Glanvill gave it legitimacy upon which Newton could later draw. In both cases, admittedly, some contemporaries bridled at being asked to accept the existence of non-mechanical forces in nature. But Glanvill's logic seemed compelling to many and Newton's to many more. Indeed, as late as 1737 William Whiston, Newton's one-time protégé and his successor to the Lucasian chair of mathematics at Cambridge University, used Newton's rationale to support Glanvill by acknowledging that if expert testimony was sufficient to prove the existence of a gravitational force, then it was equally valid as a proof of the reality of witches.

Centuries later we are far more comfortable than Newton's critics were in accepting the operation of incomprehensible forces in nature. Gravity, electro-magnetism, the strong and the weak nuclear forces are all staples of modern physics but still await comprehensive explanation. Clearly, Glanvill's argument that something can be accepted as real even if its causes aren't understood has lost none of its relevance over the last three centuries.

Passing muster

By the standards of the day, and especially those of the Royal Society, Joseph Glanvill's case for the reality of witchcraft was entirely legitimate.

Considering the widespread belief in spirits and immaterial forces, as well as the conventional assumption that Satan operated within the laws of nature, his claims weren't superstitious or far-fetched. Nor were they 'veiled in the mistie speech' of contemporary alchemy or mired in obscurantist theology. Glanvill's prose was clear and logical, his method was scrupulously rational, and it was every bit as consistent with the rubric of the Royal Society as Boyle's experiments on air pumps and Newton's theorizing about the movements of heavenly bodies. Glanvill may have begun his research already disposed to believe in witches, but he still managed to forge a sensible path between the extremes of confidence and diffidence. Only when he'd amassed a heavy weight of evidence did he go into print. And when he did, his stories demanded serious consideration.

Yet in subsequent centuries the story of the Scientific Revolution was made to fit a simple-minded, triumphalist account of science vanquishing superstition. Royal Society natural philosophers were presented as the shock troops of an intellectual revolution that owed virtually nothing to what went before. The purging of English law books during the 1730s of any mention of the crime of witchcraft was portrayed as this revolution's humanitarian fruit. Only now can we appreciate that it was entirely in keeping with the 'climate of opinion' (a phrase, incidentally, coined by Glanvill) of the Scientific Revolution that leading members of the Royal Society helped energize the belief in witchcraft just as it was beginning to recede among the educated classes more generally. Indeed, men like Boyle, Glanvill, and Newton saw themselves as the defenders of an eclectic world-view that embraced the reality of a spirit world, refuted atheism, and accepted alleged preternatural events as worthy of serious attention against the far more cynical and atheistic philosophies of London's growing number of freethinkers. What really led to the decline in the belief in magic is far from clear. It's increasingly apparent, however, that the role played by Royal Society science was far from clear-cut.

So there was no paradox in the author of *The Vanity of Dogmatising* and *Scepsis Scientifica* also writing *Saducismus Triumphatus*. Glanvill may well have been a victim of deceit and *Saducismus Triumphatus* is most certainly not the greatest example of investigative science. But, as contemporaries like Robert Boyle recognized, it certainly passed muster. On balance, Glanvill is more properly seen not as a wise fool but a missionary for Royal Society science, a man charged with such rationalist fervour that he was prepared to take the new science where others feared to tread. Or, to quote once more Robert Boyle's overtly sexualized rhetoric, Glanvill displayed a commendable willingness to explore nature's 'most intimate recesses'.

Lazzaro Spallanzani (1729–1799)

> It is impossible to convey an adequate idea of
> [Spallanzani's] many important experiments on
> generation in a few paragraphs; but it is illuminating,
> as well as regrettable, that so assiduous and
> thoughtful an experimentalist should have thought
> that he had found experimental proof for such a
> wrong theory [i.e. preformation].
>
> Arthur W. Meyer, *The Rise of Embryology* (1939).

> The old evolution [preformation] was the greatest
> error that ever obstructed the progress of our
> knowledge of development.
>
> Charles Whitman,
> *Wood's Hole Biology Lectures* (1894).

Lazzaro Spallanzani is one of the heroes of science, one of those torch-bearers who showed the way forward for his generation and whose experimental virtuosity continues to inspire others today. Yet this Italian priest-scientist has an imperfect copybook and is lucky to have kept his place among the scientific 'greats'. His grave error was to position himself on the wrong side of what's seen as one of the most misguided debates in the entire history of science. For centuries people derided medieval theologians for supposedly having debated the angelic carrying capacity of the average pinhead. But few ideas have generated such mirth or disdain as the theory, first expounded in detail by the French cleric Nicolas Malebranche in 1674, which came to be known as preformationism.

Preformationists, or ovists as they were often called, adamantly denied that children were formed from contributions made by both the

LEFT: *Portrait of Lazzaro Spallanzani.*

mother's egg and the father's sperm. Instead they insisted that each female egg contains in miniature form *all* the future generations of her lineage, each encased in the next in the style of Russian dolls. Ever since Eve, preformationists claimed, these eggs had been hatching out and would continue to do so until the original complement had been exhausted. Thus the birth of every organism in nature involved little more than tiny forms present in the woman's body from the outset expanding and filling out until they were large enough to be born into the world. Generation was, in the words of Malebranche, simply a 'kind of increase in size'.

The white fluid released by males at orgasm was thought by the ovists to be a mere catalyst, or possibly an early source of nutrition for the embryo. Either way, sperm was felt to make little or no difference to a child's attributes and abilities. Even when spermatozoa were observed for the first time in 1677, they were generally dismissed as mere parasites. A few natural philosophers, not unreasonably dubbed spermists, swung to the other extreme by claiming that the encased embryos were present in the heads of each sperm rather than in the female eggs. But the vast majority of informed opinion continued to assert that sperm merely initiated the reproductive process.

There's something profoundly demeaning about laughing at the past. It's not unlike poking fun at someone who can't speak our language and therefore seems far less intelligent than they really are. But in this instance it's very hard to imagine how acute people could honestly have believed that each egg carried the entire complement of its own posterity, nested claustrophobically one within another. From a psychological standpoint, perhaps the most surprising aspect of it was the willingness of the male profession of natural philosophy to write itself out of reproduction, a domain in which men had always seen themselves at the very least as equal partners. Yet hardest of all to understand is the preformationism of Lazzaro Spallanzani. Spallanzani was not someone who simply churned out unverifiable ideas. Here was a man whose exquisite skill at experimentation earned him the sobriquet 'Magnifico' and whose attempt to destroy the doctrine of spontaneous generation (the idea that microbes can come into existence without parent organisms), was subsequently lauded by the supreme nineteenth-century experimentalist, Louis Pasteur.

More to the point, Spallanzani performed during the 1780s a series of ingenious experiments that to us at least might seem to provide definitive proof that sperm are as important as eggs in procreation.

This chapter is devoted to exploring the reasons why such a seemingly absurd idea came to dominate so brilliant a mind. It's a story I've chosen to tell as a vehicle for showing that in the history of science truth and logic often travel apart. Understanding how even the finest eighteenth-century natural philosopher could have got it comprehensively wrong is a useful corrective for those of us who assume that the history of science is a relentless march of progress in which the best are destined to get it right. For in the period stretching from roughly 1670 till 1790, preformationism provided a far better explanation for the generation of new life than the main alternative, the 'epigenetic' or 'dual semen' model, according to which new generations are formed from material provided by *both* parents. Epigenesis won out in the end, of course, but it wasn't always the best candidate. And while Spallanzani was incorrect, this doesn't make him a bad natural philosopher or preformation a risible idea in its own time. Nor, in a just world, would Spallanzani's error have any bearing on his posthumous reputation.

Frogs, moths, and puppy-dogs

Ever humane, William Shakespeare described the destruction of flies as the sport of 'wanton boys'. In this timeless activity the young Lazzaro Spallanzani, fascinated by nature and addicted to experiment, exulted. It would have brought no comfort to the small creatures he abused, but in his case at least the motivation was philosophical inquiry rather than gratuitous cruelty. For as a child, having detached the legs and wings from insects, Spallanzani junior did his best to find means of reattaching them.

Initially, Spallanzani's interest in biology wasn't seen by his parents as promising much in the way of a career, so in his late teens they steered their son into a legal vocation. After a few months, however, he rebelled and persuaded his father to allow him to become a student of nature instead. Joining a Dominican order, Spallanzani first became a priest and for several years thereafter funded his scientific research with the income from saying masses. Thus, at the beginning, Spallanzani's religion supported his scientific activities. But such was his skill as an experimenter and lecturer that Spallanzani found himself inundated with offers from universities. He eventually took up a teaching chair at Pavia. There, in Lombardy, he assembled a menagerie of animals and set about investigating, among many other things, the profound mysteries of generation.

Spallanzani began by looking for an easy way of studying conception. He quickly saw that the simplest means by far would be to watch it occur outside an animal's body. Here, though, there seemed to be a major obstacle. According to the great Swedish naturalist, Carl Linnaeus, 'In Nature, in no case, in any living body, does fecundation or impregnation of the egg take place outside the body of the mother.' There wasn't much ambiguity there. Spallanzani, however, very much hoped that he could prove the eminent Swede wrong. And, in a series of experiments, he demonstrated that Linnaeus had indeed been too cocksure. Frog's eggs, he showed, can *only* be fertilized once they've departed from the maternal body. His method for proving this was elegantly simple. It takes in the region of an hour for a female frog's full complement of eggs to be discharged, so he took a female about to disgorge her eggs and after about half of them had been released into a tank of water containing male frogs, he removed her, extracted the remaining eggs directly from her ovaries, and then dropped these other eggs into the same tank. The results seemed unambiguous. Eggs released naturally 'became tadpoles', whereas those manually extracted from the female frog's body before their proper release time turned into 'an offensive putrid mass'.

The discovery that conception in frogs must take place outside the mother's body meant that Spallanzani could now begin to study reproduction at close quarters. This allowed him to seek an answer for a vital biological question: is male semen necessary for fertilizing the female's eggs? It might seem to us an odd question to ask, but for the committed ovist it was hard to assign any very significant role to sperm at all. Seeking to resolve this issue would absorb the next few years of Spallanzani's life.

He began by gathering together dozens of green tree frogs and 'ugly and disgusting' toads. Then he carefully watched the mating habits of his amphibious specimens. Spallanzani later described, in vivid prose, how he'd seen a female frog release her eggs as a male followed her through the water, throwing 'himself into strange contortions'. Spallanzani explained how he was struck to see that the male frog exhibited 'an obtuse tumid point', which he correctly 'suspected to be the penis, [that] now and then [was] brought towards the eggs nearest the [female's] vent'. This was highly suggestive of ejaculation. Yet no semen could be seen emerging from the frog's putative penis.

The Italian next put a male and female pair of frogs in each of several dry, empty tanks, and watched nature take its course. 'The male is so much attached to the female,' he recalled in 1785, 'that notwithstanding his being taken out of his natural element, he persists in doing his office.' Spallanzani now saw 'a small jet of limpid liquor [dart] from the small point in the vicinity of the [male's] anus, upon the eggs hanging out at the vent of the female'. Some kind of ejaculation had taken place from the pubic region of this incorrigible male. But did the ejaculate have any function? Spallanzani had to wait only a few days for his answer. The eggs that had come into contact with this liquid, and then been placed in water, 'brought forth young'. 'I hesitated not to suppose,' Spallanzani summed up, 'that the liquor emitted by the male was real semen.' This fluid, he deduced, had been necessary for fertilization to take place.

The frog tailor of Lombardy

Following up these studies, Spallanzani set about devising another, rather less naturalistic test. Following an idea first (and unsuccessfully) tried out over a decade before, he tailored pairs of underpants from a strip of taffeta fabric. These were then tightly sealed around the legs and waists of his male frogs. This is how Spallanzani described this famous experiment:

> The idea of breeches, however whimsical and ridiculous it may appear, did not displease me, and I resolved to put it into practice. The males, notwithstanding this encumbrance, seek the females with equal eagerness, and perform, as well as they

> can, the act of generation; but the event is such as may be
> expected: the eggs are never prolific, for want of having been
> bedewed with semen, which sometimes can be seen in the
> breeches in the form of drops.

Smeared onto frog-spawn using the tip of a piece of fine cord, the semen taken from inside the breeches gave life to the female's cluster of eggs. Evidence was accumulating fast. In a set of follow-up experiments, Spallanzani found aquatic salamanders to be less sartorially demanding. Male salamanders ejaculate into water before the females of the species release their eggs. So Spallanzani simply had to keep the males away from the females prior to the latter's eggs being laid. When he did so, none of the eggs hatched. But if he didn't remove the males, the water would become cloudy with semen and, several days later, the fertilized spawn gave rise to dozens of diminutive salamanders.

Spallanzani's amphibian studies had established him as a leading pioneer of artificial insemination, and he was becoming more and more aware of his experimental powers. In his next series of experiments Spallanzani 'laid open the [seminal] vesicles of dozens of frogs' and wiped their 'watery' contents with a pencil upon sets of unimpregnated eggs. When he did so, he was nearly always gratified to see the eggs 'over which the pencil had passed, begin to assume an elongated figure'. Eleven days later, tadpoles 'quitted the membranes and swam about the water.' Any eggs untouched by his pencil, in contrast, 'decomposed, and turned putrid'. It's a testament to Spallanzani's consummate skills as an experimenter that he was now able to adapt these practices to species in which fertilization occurs internally. Incredibly, he even managed to extract semen from batches of Lombardy silkworms. He then 'bathed' female silkworm eggs in the fluid he had collected, with the result that a few days later most were found to contain silkworm embryos. Eggs left untouched by semen yielded nothing.

Becoming 'even more sanguine', Spallanzani now turned to a 'bitch spaniel' in heat and 'confined her to an apartment' away from other dogs. From a 'young dog of the same breed', Spallanzani obtained 'by spontaneous emission . . . nineteen grains of seed'. These were transferred to a warmed syringe and injected into the female spaniel's uterus.

Sixty-two days later he was rewarded with the birth of 'three lively whelps'. Recognizing the importance of this experiment, he declared, 'I can truly say that I never received greater pleasure on any occasion since I cultivated natural history.'

Many of Spallanzani's contemporaries shared this assessment of his experiment's value and he won richly deserved fame. A few also realized that his methods could be applied to his own species. 'One day what you have discovered may be applied in the human species . . . with no light consequences,' an eminent correspondent observed. In fact, the first artificial insemination of a human female had already taken place. The London-based Scottish surgeon, John Hunter, had been approached in about 1776 by a draper unable to impregnate his wife because his urethra terminated on the underside of his penis. Hunter furnished him with a syringe and gave instructions for it to be kept warm when filled with semen and injected into his wife's vagina. According to papers not found until after Hunter's death, this novel technique was an emphatic success.

The mysteries of sex

Spallanzani, however, was a student of nature and not a gynaecologist. Moreover, he was also fully aware that nothing he'd done proved that semen had a unique role in the reproductive process. Might not other fluids, substances, or gases fertilize eggs equally well? Spallanzani set out to test this hypothesis with characteristic vigour. Accumulating piles of fresh frogs' eggs, he proceeded to wipe them in turn with blood from the adult heart, juices squeezed from the same organ, various chemicals, vinegar, spirits, wine (various vintages), urine, the juices of lemons and limes, and oils extracted from their skins. He even obtained a voltaic battery and passed an electric current through them. All to no avail: eventually, having exhausted his larder, garden, and imagination, Spallanzani accepted that semen is absolutely necessary for the generation of new life.

After years of experimentation with frogs, silkworms, and dogs, the Italian now seemed to have proven that semen is essential for procreation to take place. The sheer scale and rigour of his efforts denied all but the most obdurate opponents any possibility of disagreeing. But few ovists were shocked by his findings. On the contrary, one of the progenitors of the theory of preformation had explicitly recognized the role of male ejaculation. The eighteenth-century Dutch naturalist, Jan Swammerdam, had had this to say about mating frogs in 1685:

> The male frog leaps upon the Female, and when seated on her back, he fastens himself to her . . . he throws his forelegs round her breast . . . he most beautifully joins his toes between one another, in the same manner as people do their fingers at prayer . . . at last, the eggs are discharged from the female's fundament in a long stream, and the male immediately fecundifies, fertilises or impregnates them, by an effusion of semen.

But Swammerdam had denied that the semen had any structural effects upon the eggs. Instead he invoked the idea of an immaterial force, or *aura seminalis*, which was supposedly released when male organisms ejaculated. This mysterious essence, he said, stimulated the growth of the encased organism. Spallanzani was not convinced, and he set about putting the *aura seminalis* theory to a rigorous test. His most inspired experiment involved sticking dozens of frogs' eggs to the side of a glass container. He then placed inside it a small jar containing fresh frog semen. Next the vessel was sealed and left for several days, during which time the *aura seminalis*, if it existed, would have plenty of time to float up from the semen to trigger the development of the egg-bound embryos. Such experiments quickly convinced Spallanzani that the role of semen in generation had nothing to do with the actions of mysterious spirits.

Yet here comes the strange part of this story. If Spallanzani's heroic labours had demonstrated that semen is essential for fertilization to occur, nothing he saw made him so much as question his ovist position. The 'unimpregnated' frogs' eggs, he continued to assert, 'are nothing but the foetuses of the frog . . . the foetus exists in this species before the male performs the office of fecundation.' Or, as he went on to write, 'As these supposed eggs existed in the ovary before their descent through the oviducts into the uterus, and prior to fecundation, the foetus existed in the mother's body long before fecundation.'

The question we now have to ask is why, despite his wholehearted commitment to the experimental approach, his supreme skills as an experimentalist, and his prodigious intellect, Spallanzani insisted on remaining committed to what one well-informed modern historian described as, variously, the 'cloven hoof' and the 'malignant fruit' of preformationism. How could such an 'assiduous and thoughtful experimenter' have been so unambiguously wrong?

Following false trails

A little reflection tells us that at one level there's no great mystery here. If we consider the experiments Spallanzani performed in his menagerie cum laboratory, we see that not one of them cast serious doubt upon his ovist theory. His work did not yield incontrovertible evidence that male characteristics are transferred to the offspring, nor that the egg acquires form only after fertilization. All Spallanzani actually discovered is that *without* sperm procreation does not take place in certain species. And having demonstrated this, he furnished a perfectly plausible, if false, explanation as to why its involvement is so essential.

Semen, Spallanzani argued, stimulates the foetal heart into action and thereby converts an inert but preformed body into a living, breathing organism that can begin its journey along the predetermined pathway towards birth and adulthood. The male ejaculate, in other words, operates as a kind of catalyst, initiating the process of generation while not being integral to it. This idea wasn't at all unorthodox among ovists. The ovist champion, the Swiss natural philosopher, Charles Bonnet, had already come to the same conclusion: 'The interior of the Females,' Bonnet wrote, 'does not possess a Liquor subtle or active enough to open, by itself, the fibres of the Germ and thus initiate development.'

But if Spallanzani's experiments didn't disprove preformationism, neither did they embarrass its main challenger, the theory of epigenesis. His data were entirely compatible with both his own de luxe version of ovism and the epigenetic or dual-semen model. Neither was more or less plausible at the end of Spallanzani's exhaustive labours than at their beginning. The Italian, however, opted resolutely for an ovist position.

This is the real enigma of Spallanzani's scientific career. Why was his mind so firmly closed against the epigenetic accounts of generation that had been favoured by natural philosophers for millennia and, in a different form, have been taken for granted ever since the late 1700s?

To make sense of Spallanzani's unflappable faith in preformationism we need to return for a moment to the seventeenth century. Francis Bacon, one of that century's most notable scientific thinkers, referred to the province of the natural philosopher as the 'Book of Nature', contrasting it to the 'Word of God' enshrined within the Old and New Testaments. Bacon himself, however, was under no illusions as to the difficulties involved in reading nature's vast, convoluted text. The meanings of nature, he wrote, are usually 'veiled and incomprehensible'. Men, Bacon noted elsewhere, are not 'on such familiar terms with nature that in response to a casual and perfunctory salutation she would condescend to unveil for us her mysteries and bestow on us her blessings'.

Yet, when stripped of these caveats, Bacon's metaphor of the 'Book of Nature' easily misleads the unwary by seeming to imply that investigating the physical world is the direct equivalent of reading the prose of a known language. This is so far from the truth that the only way to salvage the analogy is to suppose that the Book of Nature comprises hieroglyphs. the meaning of which has to be pieced together with a combination of intuition, guesswork, speculative simulations, and exhaustive trial and error. The equivalents of the Rosetta Stone in science are great rarities. In consequence, it's all too easy for ambiguous signs to be misinterpreted and for students of nature to be led seriously astray. This, however, comes with the territory. Scientists have no alternative but to construct their world-views from fragments of evidence. In a sphere of activity constantly moving into the unknown, uncertainty is unavoidable.

Preformationism is a case in point. For far from being an absurd idea swallowed whole by a gullible scientific elite, it was a theory handsomely buttressed with evidence. But only with the benefit of hindsight can we can fully appreciate just how disingenuous this evidence was. The first false scent was picked up at the residence of the eminent French citizen, Melchisedec Thévenot, in the mid-1600s. There a usually diffident physician and microscopist, Jan Swammerdam, stunned the house guests by gently peeling off the skin of a silkworm and revealing beneath it the wings, proboscis, antennae, and legs of the future moth. An entire organism, Swammerdam declared, is wrapped up inside another, awaiting its proper time to unfold and emerge gloriously alive. Could not the same principle, he asked his elated audience, be applied to generation?

At about the same time in Paris, Nicolas Malebranche was confronted with similarly extraordinary sights of miniature organisms packed inside seeds. 'By examining the germ of a tulip bulb,' he wrote with palpable excitement, 'we discover very easily the different parts of the tulip.' But it was the chicken egg, studied up to the optical limits of his microscope by the Italian physician, Marcello Malpighi, which really seemed to clinch the case for the ovists. Malpighi accepted that copulation is necessary for eggs to be fertilized. But he mistakenly assumed that unincubated eggs that have been fertilized do *not* undergo any development. So one sees inside them, he wrongly surmised, only what already existed *before* fertilization. He was therefore amazed to see in eggs fertilized but not yet incubated embryos containing all the basic parts of chicks. 'The imprisoned animal came into view,' he later recalled, 'the stamina of the chick already exists in the egg.' Malpighi didn't explicitly declare himself in favour of ovism, but Malebranche gleefully seized on his observation: 'We can see in the germ of a fresh egg that has not yet been incubated a small chick that may be entirely formed.' How could the semen have more than a triggering function, Malebranche asked, if the fully formed chick was already present long before the cock's sperm entered the chicken's uterus?

Knowing what we now know, there has to be something wrong with this observation. The error lay in Malpighi's waiting several days after the chickens had copulated before cracking open the eggs and examining them. In the meantime, as the weather was very hot, his eggs had been incubating in the sun's heat. By the time Malpighi spread the egg yolks beneath his microscope, the development of the embryos had already advanced some way. Entirely unaware of this, Malpighi innocently published the flawed results that Malebranche seized upon.

Small worlds and asexual lice

Seeing the rudiments of a chicken in what was thought to be an unincubated egg is, of course, very different from proving that each egg contains millions of distinct embryos stored like Russian dolls. But the invention of microscopes powerful enough to reveal the hitherto invisible world of microbes gave encasement a plausibility that it had previously lacked. Nicolas Malebranche enthused that, with this newly invented instrument, 'we can easily see animals much smaller than an almost invisible grain of sand; we have seen some even a thousand times smaller . . . The imagination is overwhelmed at the sight of such extreme smallness.' The same sense of awe gave rise to Jonathan Swift's famous lines:

> So, naturalists observe, a flea
> Has smaller fleas that on him prey;
> And these have smaller still to bite 'em;
> And so proceed ad infinitum.

With every improvement in the quality of microscope lenses, ever smaller organisms were brought into view. There seemed to be no limits to the minuteness of these 'little animals', and many philosophers began quite reasonably to imagine the possibility of organisms encased one within another thousands or even millions of times over. As Malebranche wrote, 'there might be smaller and smaller animals to infinity.' From which belief it made perfect sense to argue that 'all the bodies of men and animals born until the end of times [might have been] created at the creation of the world.'

While Malebranche extolled the infinite potentialities of the world of microbes, Swammerdam unwittingly heaped error upon error in his reading of the insect evidence. When examining sliced insect larvae, Swammerdam claimed to be able to see the adult fly's 'legs, wings, trunk, horns and every other part of the animal'. This further convinced him that flies' eggs contain miniature adults that simply have to grow and unfold in order to emerge as adult organisms. What Swammerdam actually saw, however, were larval 'imaginal disks', clusters of undifferentiated cells that appear in the larvae and only gradually develop into the adult's bodily components. Unfortunately, at the level of microscopic resolution available in the 1600s, these imaginal disks looked deceptively like tiny body parts. It was a mistake just waiting to be made.

Half a century later, the humble louse charmed the ovists further into error. In 1740 the pre-eminent ovist, Charles Bonnet, discovered parthenogenesis, or asexual reproduction in animals. He showed that female lice, kept quite apart from males, still managed to produce eggs that hatched and gave rise to further generations of asexually reproducing lice. Bonnet became famous almost overnight. But he paid a heavy price for it. His philosophical colleagues began wondering if asexual reproduction was really taking place. Might one insemination, they asked, not last for several generations? Could the embryonic lice perhaps be sexually reproducing in the mother's womb?

These questions drove Bonnet to undertake a dangerously intensive bout of microscopical research. As he pored over his microscope day after day, focusing intently on a small circle of light, he reduced himself to near-blindness. By then, however, his work on parthenogenesis had done much to further the credibility of ovism by undermining the epigenetic belief in the necessity for sperm in forming new animals. Such was the perceived importance of Bonnet's work that Spallanzani became his most devoted admirer.

Having initially entered the field as an unlikely outsider, ovism could now draw great strength from hundreds of observations made by many of the most skilled microscopists of the eighteenth century. Underpinned by a growing awareness of the microbial world, and buttressed by the now indisputable fact of parthenogenesis, the idea of encasement had been rendered far from ridiculous. Ovism still had its critics, but they posed only a modest threat.

Sceptics who pointed out that miniature animals were never seen in recently laid eggs were easily dealt with. Ovists simply asserted that microscopes were not of sufficiently high resolution to reveal the tiny preformed organisms they contained. By making some modest concessions, Bonnet even found a way of accounting for such troublesome observations as the appearance in the child of its father's characteristics. His revised variant of ovism assumed that the ovum contained not a perfect miniature adult, but a slightly more basic prototype that could undergo superficial changes in response to external influences, including those exerted by the father's semen. In this manner a child could acquire, say, its father's skin colour, receding chin, facial tics, or sandy hair. But this didn't help to explain the inheritance of paternal characteristics through multiple generations of sons and daughters: how could a father's sperm alter the preformed embryos carried by his unborn daughter?

This wasn't, however, as great a problem as might be imagined. The first pedigree chart tracing the descent of a biological trait didn't appear in print until the 1750s. Even then, it was possible to see cases of apparent continuity as no more than coincidental (the equivalent of getting ten heads in a row when flipping a coin). In addition, similarities could also be accounted for in terms of the ancient doctrine that what a woman sees during her pregnancy will affect the appearance of her child. If one accepted Bonnet's idea of there being a preformed prototype, one could easily assert that traits were passed on because wives looked upon their husband's or father's features and in so doing influenced the appearance of their children. In any case, it hardly seemed reasonable to ditch a theory that explained so much in the face of difficulties which ovists felt would one day be fully reconciled to their theory anyway.

The 'mechanical philosophy'

Encased tulips, flies, moths, and chickens made ovism a popular and entirely credible theory. But this is not an adequate explanation for its ascendancy over epigenesis, the dual-semen alternative. After all, it really was a problem for the preformationist that no one had yet managed to see a rudimentary organism in an unfertilized egg. And explaining how the first rabbit's ovaries managed to accommodate $10^{10,000}$ eggs remained a daunting task. Despite these difficulties, preformationism won huge support from men like Spallanzani for one main reason: whatever the practical difficulties entailed, at the theoretical level it was graced with exquisite and elegant simplicity. Throughout the second half of the seventeenth century and most of the eighteenth, preformation provided by far the most economical explanation for one of the most inscrutable processes in the biological realm. To understand why this was so, we need to consider the intellectual milieu in which the ovists developed their ideas: the Scientific Revolution.

The point has already been made that medieval natural philosophers had despaired of properly comprehending such baffling natural phenomena as magnetism, electricity, and life. Such things seemed so inexplicable that philosophers came to believe that the best they could do was to attribute them to immaterial powers, what they called 'vital' or 'occult' forces. Different forces were said to give everything from medicines, gunpowder, magnets, and organs of the body to coffee, opium, and tobacco their mysterious properties. By invoking these occult forces, natural philosophers felt they were simply facing up to the natural limitations of the scientific endeavour. The result, however, was a proliferation of occult forces that really didn't explain anything at all.

The Scientific Revolution ushered in a period of far greater self-assurance among students of nature. To the 'new men' it seemed that earlier generations had suffered from a collective stage-fright, an overweening humility that had prevented them from properly investigating the natural world. Probably never before had a class of men had the same confidence in their capacity to understand, control, and exploit the world about them. The mindset of this period is particularly associated with the French natural philosopher, René Descartes. The universe, he argued, is really no more mysterious than an incredibly complicated clock. And like a well-made timepiece, Descartes maintained that it behaves with absolute regularity, obeying strict mechanical laws set in motion by a God who then left everything to unfold without interruption. For Descartes, the reliance of earlier natural philosophy on multiple occult forces was rank superstition. Nature involved nothing but matter in motion. Typical of the way he approached the task of explaining natural phenomena was Descartes' argument for how the tiniest particles of matter bond together. Mysterious forces of attraction didn't get a look in. For Descartes, atoms were held together by microscopic hooks and barbs.

As already noted, while Newton is now thought of as the doyen of the Scientific Revolution, some of his contemporaries considered him a dangerous reactionary. The way he upset them neatly illustrates some of the lineaments of the new philosophy. In 1687, when Newton published his theory of universal gravitation in his *Principia*, two of Europe's most brilliant thinkers, Gottfried Leibniz and Christiaan Huygens, recoiled at his idea of an immaterial force called gravity acting upon objects at a distance. For them any mention of mysterious, invisible forces was an affront to scientific progress: it was inconsistent with the Cartesian metaphor of the clockwork universe. Newton's theory seems to 'involve continuous divine intervention', Leibniz famously scoffed, rendering it a work of 'manifest stupidity', no more credible than 'tales of fairies'.

But this contempt was not felt by all. As we saw in the chapter on Joseph Glanvill and witchcraft, most natural philosophers were happy to accept the activity of some immaterial forces, so long as they weren't gratuitously invoked and were seen to obey fixed laws. Newtonian gravity passed these tests because it was a single force which explained so much, and it acted with mathematical regularity. But what of the occult forces traditionally used to explain such things as generation?

The 'man machine'

The mechanical philosopher's emphasis on physical forces and inviolable laws did not sit at all comfortably with the previous millennia of biological thought. Most of the kinds of biology that dominated the scene before the scientific revolution can be grouped under the heading 'Aristotelian'. Aristotelian biologists had made liberal use of the idea that immaterial, vital forces work upon matter to induce movement and change. As we've seen, these occult powers weren't believed to be amenable to scientific demonstration, but it was deemed essential to accept their existence if nature's stranger processes were to be explained.

Again following Aristotle, biologists had traditionally believed in some form of epigenesis, arguing that offspring are formed out of the undifferentiated mass of both parents' seminal fluids. Rather than being preformed, new organisms were believed to develop sequentially from this initially formless material. Crucially, immaterial forces were thought absolutely necessary for this epigenetic development to take place. Thus, for Aristotle the male's semen contained an immaterial potency, called pneuma, which was required to give shape and form to the female's purely material contribution to the next generation. William Harvey, discoverer of the circulation of the blood, claimed in a book of 1651 that the female egg was actually more important than the sperm. As an Aristotelian, however, Harvey was also convinced that a grand organizing force, the expression of an omnipresent God, oversaw the embryo's entire course of development from reproductive soup to actual being. But even as Harvey wrote his treatise *On the Generation of Animals*, the intellectual atmosphere was changing. His somewhat carefree allusions to immaterial forces were already making many of his contemporaries cringe.

By the mid-1600s, with the emergence of the mechanical philosophy, a growing number of biologists came to the view that they could do a lot better than simply invent a new force every time a puzzling phenomenon was observed in nature. So for the ovist Jan Swammerdam, Harvey's sweeping speculations simply illustrated the 'egregious mistakes we are apt to commit, the moment we abandon solid arguments furnished by experiments, to follow the false lights struck out by our weak and imperfect reason'. Although Swammerdam himself had stipulated the existence of an immaterial *aura seminalis*, he felt that this was small beer in comparison to the epigenesist's grand organizing forces. Having taken in deep draughts of mechanical philosophy, he was determined to replace epigenesis with a much less mystical model of generation. 'For all the foundations of all created beings are few and simple,' he wrote in a book of 1685, 'so the agreement between them is most surprisingly regular and harmonious.' Epigenesis made no sense at all from this perspective because turning the 'semen' of both parents into a complete animal would seem to require a whole array of mysterious forces. Epigenesis, therefore, had to be wrong.

Nicholas Malebranche too was inspired by the new trend for mechanistic thinking and the motif of the clockwork universe. At the age of 27 he read *L'Homme de René Descartes* and was so excited by it that he reputedly experienced heart palpitations. In later years he rowed back to some extent by conferring on God a more active role in the operations of the universe. But he still repudiated Harvey's vital epigenetic forces.

The mechanical philosophy, however, did not lead Swammerdam and Malebranche, or for that matter Bonnet and Spallanzani, to preformationism by a direct route. The most obvious choice for the new wave of philosophers was to construct a mechanical model of epigenetic generation. René Descartes had tried to do just this towards the middle of the seventeenth century. 'It is no less natural for a clock,' he argued, 'made of a certain number of wheels, to indicate the hours, than for a tree from a certain seed to produce a particular fruit.' But this clockwork analogy seemed a little too simple-minded even for Descartes' most ardent admirers. 'All the laws of motion which are as yet discovered,' wrote the philosopher George Garden in 1691, 'can give but a lame account of the forming of a plant or animal. We see how wretchedly Descartes came off.' Spallanzani fully agreed. 'It is incomprehensible,' he wrote, 'how an unorganized body of solid and liquid form, could possibly be organized by means of mechanical laws alone.'

This left only one path open. On the one hand, immaterial forces had to be kept to a minimum according to the ethos of the mechanical philosophy, while on the other, Cartesian matter in motion patently wasn't enough to account for generation. Bereft of alternatives, biologists were driven into the arms of preformation. As they recognized, the problem of explaining how reproductive fluids turn into babies simply disappears if it's assumed that the baby already exists, preformed, in the egg. At this stage, what we now know as Sherlock Holmes' dictum came to apply: 'When you have eliminated the impossible, whatever remains, however improbable, must be the truth.' The mistake Spallanzani and his colleagues made was to think epigenesis impossible; but having made this entirely understandable error, it was perfectly reasonable for them to assume that ovism had to be the answer.

Nevertheless, biologists like Spallanzani embraced ovism willingly and with their eyes wide open. For quite apart from the physical evidence accumulated by Swammderdam, Malebranche, Malpighi, and later Bonnet, ovism made a great deal of theoretical sense.

God's infinite wisdom

With very few exceptions, both ovists and epigenesists were in agreement that God had had the power to do anything He wished when creating the universe. Claiming that He had stocked the loins of Eve and the progenitors of all other species with enough miniature organisms to last till Judgement Day presented no serious intellectual difficulties. After all, in comparison with the wondrous plenitude of animal and plant life, the extraordinary richness of the microscopic world, and the seemingly infinite number of stars in the heavens, there was no reason to think that encasing even millions of eggs within eggs would have seriously taxed the Creator's ingenuity. In an age in which the word God implied unlimited powers, ovism was in no sense far-fetched. As the modern, South-African born biologist, Lewis Wolpert, has observed, 'such a thought seemed extravagant only to those who measured God's powers by their own imagination.'

Even though Descartes' influence had waned somewhat by the later 1700s, for natural philosophers like Spallanzani the issue remained that of choosing between a model of generation that required grand immaterial forces and one that did not. For most, there was simply no contest. If all the preformed embryo had to do was get bigger, then generation could be brought into line with the same rationalist approach that structured the period's new scientific world-view. By denying that embryonic development actually occurred, ovism cleverly sidestepped all the problems confronted by its rivals. So, rather than having to infer the actions of lots of elaborate, vital forces, philosophers turned instead to the comparatively intelligible idea of encasement.

It was the apparent rationalism of ovism, then, that gained it so much popularity. The French naturalist, Claude Perrault, for instance, wrote in 1688 that preformation had to be right because it was 'inconceivable' that a complex being could 'form itself from matter out of chaos'. Albrecht von Haller, among the greatest medical writers of the eighteenth century, converted to ovism from an epigenetic position because he too realized that it offered a way around having to invent a vast array of occult forces. As he elegantly explained in 1752:

> If the first rudiment of the foetus is in the mother, if it has been built in the egg, and has been completed to such a point that it needs only to receive nourishment to grow from this, the greatest difficulty in building this most artistic structure from brute matter is solved. In the hypothesis, the Creator himself,

> for whom nothing is difficult, has built the structure: He has
> arranged at one time . . . the brute matter according to foreseen
> ends and according to a model preformed by his Wisdom.

Bonnet made much the same point. For him, ovism disposed of the metaphysics he felt had hampered progress in biology since the Greeks. Hence he saw it as 'one of the greatest triumphs of rational over sensual conviction'.

During the mid-1700s a few natural philosophers tried to rehabilitate epigenesis. Most of them argued that new beings were produced from hereditary material donated by both parents coming together in accordance with the Newtonian law of attraction. Yet while few denied Newton's genius, to most people basic laws like attraction were no more adequate in accounting for the generation of new life than Descartes' simplistic clock analogy. As one critic of this view put it, 'I do not find in all nature the force that would be sufficiently wise to join together the single parts of the millions and millions of vessels, nerves, fibres, and bones of a body according to an eternal plan.' Preformationism remained the only decent solution to the problem of generation.

Against this background we can fully understand why Lazzaro Spallanzani stuck with ovism despite all his experimental findings. He could happily accept the linear simplicities of gravity; but he had no truck with the kinds of complex immaterial forces that seemed necessary for epigenesis to be true. And so, he wrote in 1784, '[e]loquence and cleverness will never lead us to believe than an animal or plant could originate from an accumulation of parts.' In the absence of a viable epigenetic alternative, he resolutely and entirely sensibly embraced ovism.

A new paradigm

Yet for all the cogency of Lazzaro Spallanzani's arguments, he was to be the last distinguished upholder of the doctrine of preformation. Within a few years of his death in 1799 the theory had been comprehensively routed by new variants of epigenesis. The long reign of ovism was abruptly terminated and preformationism fast became an embarrassing memory, best repressed. This rapid demise had many causes, cultural and scientific.

Secularizing trends that inspired the belief that God had little role in the day-to-day running of the universe advanced to the point at which ovists were thought to be laying far too much emphasis upon the ingenuity of the Creator. Although rarely atheists, men of science in the early 1800s were more and more committed to giving all natural phenomena genuinely physical explanations. Indicative of this shift, geologists now agreed that the question of the age of the Earth should be settled by rocks and fossils rather than by totting up the number of human generations since Adam and Eve, as documented in the Old Testament. In such a milieu, even to the limited extent proposed by the ovists, it was no longer acceptable to place the divine at the heart of scientific explanation.

The revival of epigenesis was also assisted by the emergence of a new 'vitalistic' paradigm in the late 1700s. This held that dynamic forces were properties of matter itself and not the result of forces acting from outside according to the dictates of an immaterial soul. Germ cells, for instance, were said to have the intrinsic power to grow and develop. Stomach cells had the inherent power to digest. Brain cells had the in-built power to think and remember. These internalized forces were deemed to be fundamental to living tissue, and once they were dissipated or destroyed, death inevitably ensued. Belief in such forces was called materialism, and although it had plenty of historical precedents, its impact on biology grew rapidly during the early decades of the nineteenth century. Materialist biologists concluded that machine metaphors weren't very helpful in explaining bodily functions and that there really is something distinct about living matter. They had no conception of the role or even the existence of genes; they knew little about the complex electro-chemical processes that make life possible; but they had arrived at the important supposition that whatever principles do determine growth, reproduction, digestion, cognition, and so forth, they are material properties of the body itself. Biology's love affair with preformationism was at an end.

I think we've seen enough to appreciate that the eventual abandonment of ovism does not mean that the idea of preformation had been a foolish diversion that arrested the development of biology. In fact, the ovists performed a vital role by helping to rid biology of the idea that the human body is controlled by all manner of mystical forces operating from outside it. Ovism helped prepare biology for materialism by reducing the role of external forces down to a single divine act. In doing so, its exponents provided a crucial stepping-stone, first to what might be termed neo-epigenesis, and then on to modern genetic theory. Put another way, by repudiating most forms of vitalism the ovists can be seen to have broken the stranglehold of Aristotelian biology and thus made possible the advances of nineteenth-century life science. One thing then is clear: the period in which the preformationism of Spallanzani ruled supreme was very far from being a dark age of biological thought.

But it would be unacceptably presentist to wipe Spallanzani's copybook clean simply because he inadvertently helped biology along the road to modern orthodoxy. In any case, we have another justification for rehabilitating him. For Spallanzani very properly adhered to a theory that at the time was not only more plausible than anything else on offer, but made far better sense of the available data. With the help of Bonnet and others, he melded theory and experimental evidence into a unified thesis of outstanding conceptual quality. Whatever list of criteria we might devise to identify good science, Lazzaro Spallanzani's would surely qualify.

Max von Pettenkofer (1818–1901) and Robert Koch (1843–1910)

> Koch sent him a tube that swarmed with wee
> virulent comma microbes. And so Pettenkofer . . .
> swallowed the entire contents of the tube. . . and
> said: 'Now let us see if I get cholera!' . . . the failure of
> the mad Pettenkofer to come down with cholera
> remains to this day an enigma.
>
> Paul de Krieff, The Microbe Hunters (1926).

In 1892 the Bavarian chemist, Max von Pettenkofer, wrote a letter to Robert Koch, a brilliant bacteriologist who in 1883 had discovered the micro-organism that causes cholera. In his letter Pettenkofer asked Koch to send him a flask containing a sample of cholera germs. Koch obliged, supposedly obtaining fresh cholera 'vibrios' from a victim of the epidemic then raging in the city of Hamburg. Having received Koch's batch of deadly germs, Max von Pettenkofer diluted them in a flask of meaty gravy and then drank the lot. Sometimes fortune does favour the brave. This dose of germs should have killed Pettenkofer. But, remarkably, he escaped with only a brief spell of diarrhoea. The experience left him exultant. While his actions might smack to us of suicidal idiocy, Pettenkofer himself was absolutely convinced that bacteria alone are never enough to induce cholera. So, having settled his bowels, he sent the following message to Koch:

> Herr Doctor Pettenkofer presents his compliments to Herr
> Doctor Professor Koch and thanks him for the flask containing
> the so-called cholera vibrios, which he was kind enough to
> send. Herr Doctor Pettenkofer has now drunk the entire
> contents and is happy to be able to inform Herr Doctor
> Professor Koch that he remains in his usual good health.

LEFT: *Max von Pettenkofer © Science Photo Library.* RIGHT: *Robert Koch © Hulton Archive/Getty Images.*

Despite the flippant tone of his letter to Koch, Pettenkofer believed that he'd performed an act of rare heroism. 'Mine would have been no foolish or cowardly suicide,' he later observed, 'I would have died in the service of science like a soldier in the field of honour.'

The modern medical scientist would almost certainly agree that self-experimentation ranks among the highest expressions of scientific heroism. For injecting himself with syphilitic matter, charting the disease's course, and then describing the supposed curative powers of ingesting mercury, the eighteenth-century Scottish surgeon, John Hunter, has an assured place in the medical honours list. So does another Scottish surgeon, James Young Simpson. In the mid-1800s he championed the surgical use of the anaesthetic chloroform after personally testing its numbing effects with a group of friends in Edinburgh. Such uncommon disregard for personal welfare in pursuit of medical advance provides us with a real-world, secular equivalent of the romantic knight's quest. But heroism alone is not enough to ensure posthumous glory. Science is much more about uncovering the truth than self-sacrifice. And self-experimenters who are not seen to have advanced the cause of human knowledge are fortunate if they are simply forgotten. Max von Pettenkofer has not been so lucky.

Kindlier historians have characterized Pettenkofer as a tragicomic figure. The popular medical writer, Paul de Krieff, for instance, provided a vivid portrait of 'mad Pettenkofer' as an eccentric but well-meaning blunderer. Others have imputed mental instability to him on the basis that only a madman would have taken so absurd a risk with his own life. The fact that in February 1901 Pettenkofer pressed a revolver to his right temple and blew his own brains out is often used as supporting evidence for this diagnosis. But on one issue nearly all historians of the period used to agree: Pettenkofer's survival in 1892 delayed the ultimate success of the hygiene movement and thereby cost lives. For this reason, some have preferred to cast him as a shameful aberration. The twentieth-century historian and public-health reformer, Sir Arthur Newsholme, for instance, laid at his feet much of the responsibility for the 'inadequate emphasis' placed upon achieving a 'perfectly safe water supply' by the early sanitarians. Pettenkofer's influence, Newsholme implied, was more than quixotic: it was deadly.

Whether he's seen as mad, bad, or just plain stupid, Pettenkofer's reputation has suffered grievously over the last hundred years. In the long term, his flask of cholera vibrios became, figuratively as well as literally, a poisoned chalice. The reason for this descent into notoriety couldn't be more straightforward. Pettenkofer had the effrontery to deny that dangerous micro-organisms being passed directly from person to person can spread disease. He refused to accept what everyone now knows to be true: that bacteria transferred directly from one body to another can induce sickness in the new host. Worst of all for Pettenkofer's standing in the halls of science, he persisted with this claim just when the great heroes of nineteenth-century medical science, Louis Pasteur and Robert Koch, were formulating the modern germ theory of disease.

In this chapter, however, I argue that history's verdict has done little justice to a man once regarded as Bavaria's greatest gift to science. In challenging Pasteur and Koch, Pettenkofer was doing exactly what any textbook of scientific practice would have told him to do. He first highlighted the flaws in his opponents' ideas and then formulated an alternative theory that made excellent sense of data that his arch-rival, Robert Koch, simply chose to ignore. Swallowing a flask of cholera vibrios was the act of a genuinely brave man who felt that only such a grandiloquent gesture would make his opponents think again. As we'll see, the ultimate robustness of modern germ theory owes much to the fact that some scientists, if not Koch himself, were paying attention.

The X, Y, and Z of epidemics

If we're to understand Pettenkofer's ideas about germs and epidemics we need to appreciate the broad contours of medical opinion in the 1850s, the decade in which he joined the fray. Until the 1870s germ theory was an interesting speculation, but no more. Back in 1850, Louis Pasteur was an ambitious chemist in Lille still to win his scientific spurs, and Robert Koch just a precocious boy growing up in a mining town in a mountainous region of eastern Germany.

The then prevailing paradigm held that epidemic diseases were caused by poisonous fumes, called miasmas, emanating from the decomposing animal flesh, effluents, cesspools, and rotting vegetable matter that littered the poorest and unhealthiest parts of industrializing Europe. Before anything was known about the role of germs in causing sickness, this association between malodorous air and disease made complete sense. After all, epidemic diseases do favour conditions characterized by the public-health reformer, Sir John Simon, as 'excrement-sodden earth, excrement-reeking air, and excrement-tainted water'. But general agreement on miasma theory didn't prevent major divisions in other areas.

One major point of contention concerned the way in which epidemic diseases spread. At one extreme end of the debate lay the 'localists'. They argued that each and every case of a disease was due to an individual's direct exposure to miasmatic particles. Only by eating the same contaminated food or inhaling the same deadly bouquet would another person fall victim to the disease. Unless the source of the poison was transferred, there was no danger of an epidemic being spread from person to person or from town to town. In contrast, 'contagionists' insisted that individuals with the disease become infectious, releasing the seeds of further outbreaks in their breath, sweat, stools, or upon their skin, clothes, and possessions. As one might expect, most doctors positioned themselves somewhere between these two extremes, with many arguing that epidemics start with airborne miasmas but are then propagated through direct contact with victims. Perhaps typical of the man, Pettenkofer entered the debate by offering his own distinctive compromise.

He began by rejecting miasma theory and embracing the highly controversial idea that specific germs are responsible for causing epidemics. In the case of cholera he argued that a distinct micro-organism does the damage and that it's released in the victim's watery stools. This was a bold position to adopt thirty years before Koch identified the comma-shaped bacillus responsible for the disease. Less positively, Pettenkofer's early allegiance to the idea of specific germs was not based on overwhelming evidence. It was also inextricably tied up with a set of ideas that no one would call prescient.

Pettenkofer anticipated Koch by claiming that it was impossible for someone to fall sick with cholera unless the hypothetical germ, called 'x', was present in their intestines. But unlike Koch, he believed that although 'x' was a necessary cause of cholera, it was not a sufficient one. In order for 'x' to cause illness, he argued that it first had to 'ripen' in an appropriate medium, which he called 'y'. The best form of 'y', Pettenkofer said, comprises moist soil, neither too sodden nor too parched, and containing plenty of rotting organic matter such as human or animal effluents. Once bathed in 'y' for a few days, the cholera germs, 'x', matured and began producing a dangerous substance called 'z'. Only having consumed food, drunk water, or inhaled air containing a germ in this stage would the victim's intestines come into contact with 'z' and full-blown cholera ensue. Crucially, Pettenkofer was a localist to the extent that he did *not* believe in direct person-to-person contagion. Having passed through one victim's body a germ had then to spend some time in an appropriate soil medium before it could become infectious once more.

These are ideas we now know to be wrong, but they had some very positive implications for public-health policy. Although Pettenkofer's ideas were taken to imply that quarantine measures were futile, they did focus attention on the importance of improving sanitation and water supplies. Fortunately, several German towns and cities were taking careful note. Citing Pettenkofer's germ-soil theory, Lübeck, Berlin, and Munich all undertook extensive sanitation reforms. Munich, Pettenkofer's hometown, was especially pro-active. The ground beneath its streets and houses was a honeycomb of old and new sewer pipes, pump reservoirs, privy vaults, and shallow wells. Taking Pettenkofer's advice, the city authorities embarked on a large-scale building programme, designed to keep the city's water supply free from contaminated soil and prevent organic waste from providing the mulch supposedly needed by ripening cholera germs. Pure mountain water was piped into every house, and a complementary sewage system channelled effluent into the fast-flowing River Isar, which rapidly carried it away.

These reforms were thoroughly beneficial. Munich, writes one historian of public health, 'became one of the healthiest cities in Europe, thanks to the efforts of this energetic hygienist'. This extraordinary success story gives the lie to the claim that Max von Pettenkofer set the public-health movement back decades because he didn't care about clean water.

Evidently, in the world of hygiene flawed ideas aren't always bad news. But even scholars who recognize Pettenkofer's practical achievements condemn the theorizing that underpinned them. Dozens of accounts tell of how Koch systematically destroyed Pettenkofer's arguments but that, like an impulsive child, he wouldn't desist from expounding them. To assess the fairness of this portrayal we need next to look at what Koch was claiming and the basis upon which he did so.

In a golden decade, starting in 1873, Robert Koch rose from the relative obscurity of general practice to become one of the world's most respected medical scientists. By way of a series of elegant and determined studies, he probably did more than any other individual—Pasteur included—to give germ theory scientific respectability. His crowning achievement during that period was the identification of the bacteria responsible for tuberculosis.

For centuries this terrible killer had been blamed on bad heredity or fragile constitutions. Undaunted, Koch hunted relentlessly for a microbial cause and he eventually found it. But convincing doctors to abandon what they'd always believed was no simple matter. So Koch began by publicly setting out what was logically required in order for everyone to deem his claims true. Now known as Koch's postulates, they comprise four conditions that have to be met before it can be said with confidence that a specific germ causes a specific disease: first, the bacterium must be present in every case of the disease; second, the bacterium must be isolated from the diseased host and grown in pure culture; third, the specific disease must be reproduced when a pure culture of the bacterium is inoculated into a healthy, susceptible host; and fourth, the bacterium must be recoverable from experimentally infected hosts. Brilliantly fulfilling all four postulates, Koch's tuberculosis research was heralded as a genuine breakthrough.

In 1883 a triumphant Robert Koch then turned his attention to another of the horrors of the Victorian age: cholera. Europeans had enjoyed a decade's respite from this terrible disease when news arrived that Egypt was in the grip of a severe outbreak. About 5000 deaths were being recorded each week. To localists this posed no serious threat to Europe because, in their view, environmental conditions were so different in North Africa that there was no significant risk of the disease spreading.

Contagionists like Koch correctly surmised, however, that cholera had arrived in Egypt on a ship travelling through the Suez Canal from India and would soon continue its journey to Europe and America.

The British were acutely sensitive to any suggestion that cholera had been imported from India. Having taken effective control of Egypt in 1881, the government in Westminster had rescinded almost all quarantine procedures in the ports of Cairo and Alexandria. This had been done for purely commercial reasons. Quarantine reduced the speed and hence the profitability of the vast British commercial fleet that travelled through the Suez Canal *en route* to and from India and the Far East. Suddenly alarmed that it might incur the wrath of the Egyptians and the odium of the World, the British government quickly assembled and despatched a team of scientists to Egypt to investigate the problem. Knowing the answer they wanted, care was taken to pack this group with committed localists, strongly opposed to the contagionist germ theories of Pasteur and Koch. True to form, after a few weeks in Egypt they concluded that weather conditions there had 'reactivated' cholera poisons that had lain dormant in the sandy soil since 1865. Their sponsors hoped that these findings would quell fears of the epidemic's spread and exonerate the British government from all blame.

Koch's expedition arrived in the Egyptian capital shortly afterwards. Having dissected the corpses of as many cholera victims as they could find, his team encountered large numbers of a curious comma-shaped bacillus in the intestines and bowels of their cadavers. Running out of bodies to dissect as the epidemic died down, Koch's men loaded their animals and equipment onto a ship bound for India and continued their hunt there.

In Calcutta, they again found their comma-bacillus in the lower intestines of cholera victims. Koch wired news of this finding back to Germany. 'It can now be taken as conclusive,' he wrote, 'that the bacillus found in the intestine of cholera patients is indeed the cholera pathogen.' By the close of this expedition, Robert Koch believed that he'd discredited both the standard miasmatist argument and Pettenkofer's germ-based refinement. But Koch knew from the outset that his assertions would be disputed by Pettenkofer, and it gives us some sense of the older man's importance that Koch interrupted his return journey from India to Berlin in order to visit Pettenkofer in Munich. If he'd hoped to make a convert of him, Koch must have left disappointed. Pettenkofer received the younger man's claims with predictable ambivalence. He was happy to accept that someone had at last found a specific cholera microbe. But he steadfastly refused to accept that Koch had invalidated the germ-soil hypothesis.

A rock and a hard place

Dubbed 'Father of the Bacillus' by a German press delighted with his bravura performance in Cairo and Calcutta, Koch arrived back in Berlin on 2 May 1884, a national hero. He was immediately granted an audience with the aged Kaiser Wilhelm I, who gave him a life-size bust of himself. Wilhelm was thrilled with Koch's findings, not least because of the political and commercial embarrassment his discovery had caused the British. No less enthused at Britain's discomfort, Otto von Bismarck personally congratulated Koch and gave him a specially minted medal. Not to be outdone, the Reichstag awarded him and his team 100 000 marks in gold coins. A few days later a banquet was thrown in his honour, attended by guests drawn from the highest strata of German society.

Yet once the glare of publicity had dimmed, it became clear that Pettenkofer had plenty of fight left in him and plenty to fight for. Koch's findings had not been universally well received. Pasteur led the doubters in asserting that Koch had not found the true microbial culprit. Soon after, a leading British biologist referred dismissively to Koch's Indian expedition as an 'unfortunate fiasco'. No doubt Koch interpreted this snub from Britain in the same fashion as some modern historians have done: as an expression of the calculated British aversion to any theory of epidemic disease that threatened to reduce the profits of its commercial shipping by implying the need for quarantine.

But writing off Koch's detractors as a chorus of selfish ignoramuses does not fit well with the evidence. Many of his British critics were perfectly able researchers, remote from commercial pressures. Nor was the criticism determined solely by nationality. Koch encountered nearly as many objections from his own countrymen. Cynical commercial interests were certainly in play, but there were also valid scientific reasons why Koch found himself so embattled; for, by the standards he had himself established in 1881, the evidence he'd presented in 1884 remained hardly more than suggestive.

Philosophers of science tell us that the most effective way for one theory to supplant another is by the carrying out of a single, crucial experiment that shows conclusively which theory better fits the facts. Unfortunately, as several chapters in this book indicate, such clear-cut resolutions of scientific debates are difficult to bring about. Koch's tuberculosis research did meet the exacting standards set by philosophers. But although Koch would never fully admit it, the first phase of his work on cholera came nowhere near such a conclusive result.

Having found, isolated, and cultivated his comma-bacillus, he'd had no success whatsoever in his attempts to induce the disease in the rabbits, guinea pigs, cattle, and fowls into which he injected his microbial solution. No matter how much or where he syringed his germ-laden fluid, the animals remained robustly healthy. Koch and his team concluded that man must be uniquely susceptible to the cholera germ. But, as Koch realized full well when he drew up his four postulates, claims like this had to be proven, not just made. Indeed, he'd introduced his postulates to prevent just this kind of circular reasoning.

There were at least three explanations for Koch's troubles, each of which made perfect sense of the available data: first, animals really are immune to cholera; second, the bacterium Koch had isolated was not the true cause; third, he had the right bacterium but those samples he injected into laboratory animals had not had a chance to ripen in moist soil, as Pettenkofer believed essential. In theory, Koch could have resolved the issue by administering potentially deadly injections to healthy men and women. If germs taken directly from culture plates could be shown to cause cholera in people, then he would destroy both Pettenkofer and the remaining miasmatists all in one go.

But the scientific community of Koch's day would no more have tolerated human vivisection of this kind than would modern scientists. Self-sacrifice is one thing, sacrificing others quite another. Nevertheless, knowing what he wanted to find, Koch cavalierly opted for the first of our three possibilities with few serious qualms. As the historian Richard Evans has put it, his confident pronouncement in 1884 that he'd discovered the cholera germ rested 'on the assumption that normal procedures did not apply'. In effect, Koch was denying the relevance of his own postulates. To his credit, however, Koch did recognize that he had a problem. And he fell back upon the only piece of evidence he really had: the fulfilment of postulate number one, or a 'constant association' between a person having died of cholera and their intestines containing large quantities of the comma-shaped bacillus. Yet this on its own offered no proof of causation. The comma-bacillus, it could be argued, might instead be a variety of microbe that colonizes the already infected tissues of cholera victims, much as some species of fungi thrive on already decaying wood or as scavenger animals feed on carrion.

For people who'd been impressed by Koch's work on tuberculosis but were perturbed by his failure to satisfy his four postulates in the case of cholera, Pettenkofer's theory now offered a neat and perfectly logical way out of trouble. All one had to assume was that germs were not always able to induce illness. In particular, those cultivated in Petri dishes or extracted from the corpses of cholera victims were not infectious unless they'd first spent a period of time in an appropriate soil medium. According to this interpretation, Koch's animal experiments were bound to fail even though he'd isolated and cultivated exactly the right species of micro-organism. In short, Max von Pettenkofer's theory comfortably explained all available observations.

Koch's difficulties were eventually resolved in 1895, when the Russian émigré scientist, Waldemar Haffkine, working in Pasteur's laboratory in Paris, managed to produce a strain of the comma-bacillus so potent that it overwhelmed the defences even of hitherto immune animals. Koch finally had the proof of contagiousness he had sought for so long. Over the following years Haffkine even managed to use weakened forms of cholera bacteria to produce effective human vaccines. These successes finally put paid to Pettenkofer's stand, for the bacilli used by Haffkine to infect guinea pigs and prepare vaccines had had no contact whatsoever with moistened soil. Prior to Haffkine's intervention, however, this was no more than the stuff of Koch's dreams.

Koch's blind spot

Koch tried to outflank his critics by turning to epidemiology. In the event, this only got him into further trouble. In India he located a village that had been hit hard by cholera and he set about looking for its source. He soon tracked it down to a large water tank that the locals used not only for drinking and bathing, but also for washing clothes, including the soiled garments of cholera victims. In this deadly reservoir Koch was gratified to find large quantities of his comma-shaped bacillus. But this was not the Eureka moment it's said to have been. While almost all the villagers drank water from the polluted tank, there were only seventeen confirmed fatalities. Not only did laboratory animals fail to develop cholera: many humans who ingested the comma-bacillus also remained in good health.

We now know that certain villagers survived because some people mount effective immune responses to pathogens that prove lethal in others. Some, most famously New York's 'Typhoid Mary', can even remain perfectly healthy while the disease germs use their bodies as breeding grounds. Koch was well aware that people are not equally susceptible to disease. His response was to surmise that the villagers who hadn't fallen ill had either drunk less of the water or were not predisposed to develop cholera. Beyond that, he was largely silent. For him, the seventeen victims of the contaminated Indian water tank were of overriding interest; the lucky survivors simply faded from view. But to those who focused on the survivors rather than the victims, Koch's theory seemed utterly inadequate. The evidence for contagion just didn't stand up to epidemiological scrutiny. Scientists in India were particularly sceptical because they knew of hundreds of cases in which the doctors, nurses, and families of cholera victims survived unscathed. As one report elaborated, even most of those responsible for 'removing the evacuations of cholera patients' remained fit and well.

Looking back from a better-informed present, we can see that Koch and his opponents were immersed within a paradigm that had only the haziest notions of immunity. This theoretical deficiency meant that Koch found himself under attack from scores of doctors who justifiably felt that their epidemiological data made a nonsense of his claims. For his part, Koch acquired what turned out to be a useful blind spot, allowing him to ignore lots of seemingly compelling evidence indicating that his theory was nowhere near as persuasive as he thought. But lacking his vested interest, others offered rival explanations for why some people do and some people do not contract epidemic diseases. Pettenkofer's germ-soil theory was among the leading contenders.

Reflecting on Koch's water-tank argument, Pettenkofer could simply assert that not all the cholera vibrios drunk by the Indian villagers had had a chance to mature in a proper soil medium. Those bacilli that had fully 'ripened' produced quantities of 'z' and killed the seventeen victims. Conversely, those bacilli that had come straight from the intestines of cholera victims had had little or no effect. This interpretation was a perfectly good stab at explaining an epidemiologically complex picture that Koch largely ignored. And it was in this context that Pettenkofer gulped down his now infamous flask of cholera-infected soup. He was not alone in making this bold gesture. Four of his students, Rudolf Emmerich, Elie Metchnikoff, and doctors Stricker and Ferran, also drank solutions of the cholera vibrio. It seems likely, in retrospect, that Pettenkofer and his team had been exposed to weak strains of cholera in the past and had thereby acquired immunity. Unsurprisingly, given the state of knowledge during the 1880s, Koch never really considered this possibility. Instead he rubbished Pettenkofer's auto-experiment as meaningless theatre without having any compelling basis on which to do so.

Were history written by impartial observers rather than the eventual victors, the laurels in this contest might well have been awarded to Pettenkofer. Wrong he may have been, but he had good reasons for thinking he was right.

Troubled waters

Pettenkofer's best weapons came from the masses of epidemiological data he collected during his scientific career. In a series of books and articles he showed that the incidence of cholera and typhoid does not fit the standard contagionist's model. Epidemics do not always follow trade routes closely, even though travellers are 'notoriously careless' in the disposal of their excreta. In addition, some regions along major roads and rivers seem to escape epidemics over and over again whereas other places are hit hard every time.

Pettenkofer asked himself if this variegated pattern of disease incidence could be correlated with local conditions. He began by obtaining a military map of Bavaria. He then built up a list of the addresses of all those who'd died in previous epidemics and colour-coded each street according to the number of fatalities. Something immediately stood out. Consistently, deaths were most common in moist, but not waterlogged, low-lying areas. Nor was this trend limited to his home state. In Nuremberg, in spite of the free passage of human and animal traffic from one side to the other, five times as many people lost their lives in the part of the city built on moist, sandy soil than in the dry, stony section.

Pettenkofer tried to confirm these preliminary findings during the 1850s by sinking large numbers of hollow pipes into the soil in various parts of Munich during different times of the year. His goal was to measure the ground's water content. Having collected these data, he then plotted groundwater levels against the number of cholera deaths on a chart. The results lent much support for what he'd deduced from the Nuremberg and other outbreaks: when soil is very wet relatively few cholera deaths occur. He inferred that this was because 'x' can no more prosper in sodden ground than can the seeds of wheat or barley. As he expected, Pettenkofer found the largest number of deaths coinciding with the period of lowest rainfall, when the Munich soil was moist but not waterlogged. This relationship also held for the cholera epidemics of 1836 and 1854. Only in damp soil, Pettenkofer reasoned, can bacilli switch into attack mode.

Never one for half-measures, he next challenged the medical fraternity to show him evidence of a single major outbreak of cholera occurring where houses were built upon arid soil or rock. Several took up the challenge. And far from being the slippery obscurantist of standard histories, Pettenkofer assiduously followed up each and every claim. A Professor Drasche of Vienna consulted a geological map and pointed to an area of the Karst Mountains of Croatia that had both cholera and rocky ground. Without delay, Pettenkofer went to the stricken villages and showed that the geological maps were wrong. He not only found plenty of porous soil, but also demonstrated that nearly all the cholera deaths had occurred among households built upon it. When Koch himself argued that severe epidemics had occurred in Bombay and Genoa where the soil was hard and compact, Pettenkofer was able to show that his adversary had consulted out-of-date maps. A sanitary engineer soon confirmed that the soils in both areas were moist *and* rank with rotting, organic matter: precisely the conditions Pettenkofer believed necessary for epidemics to occur.

It was no coincidence that Pettenkofer found this relationship between soil type and the incidence of cholera. On the contrary, it's precisely what modern epidemiologists would expect. In the 1850s Pettenkofer discovered a genuine epidemiological phenomenon: cholera epidemics are often associated with foul, moist soil. For reasons that aren't entirely clear but have to do with soil chemistry, this kind of medium is particularly conducive to the multiplication of cholera vibrios. And this isn't the only reason to expect a correlation between rainfall and the intensity of cholera outbreaks. Epidemics of this disease often began when water from polluted pools leaked into pump reservoirs supplying drinking water. This was far less likely to happen where the soil was dry and non-porous because rainwater drained away before having a chance to collect into pools. Very wet months of the year also saw downturns in the frequency of cholera as heavy rainfall diluted any bacteria in polluted canals, reservoirs, rivers, and streams, and thereby made the water much less dangerous to drink.

So, although ultimately proved bogus, Pettenkofer's germ-soil theory was built upon data that had somehow to be explained. In this sense Koch's greatest strength was also his greatest weakness. His remorseless focus on finding the biological origins of infectious diseases brought success unparalleled even by Louis Pasteur; but it also made him insensitive to the role played by factors such as groundwater in the emergence and collapse of epidemics. While Pettenkofer attached immense significance to the variations in disease incidence across time and space, these data were of little interest to Koch. By turning a blind eye to them, he managed to avoid getting bogged down in irresolvable complications. But in fairness to Pettenkofer, we need to recognize that he alone produced what all his colleagues strived for and what Koch was unable to deliver: a unified theory of epidemic disease.

Hail the conquering hero

In the 1880s Koch's contagionist theory of cholera was neither obviously correct nor demonstrably superior to the germ-soil alternative. Yet by the late 1870s Pettenkofer's star was on the wane. Long before Haffekine's breakthroughs of the mid-1890s, Koch had taken Pettenkofer's position at the forefront of German medical science. Koch's brilliant research was certainly a factor here. But it's not enough to explain the ease with which he put Pettenkofer in the shade. A less obvious reason for his success is that while Koch enjoyed the emphatic support of Wilhelm I and Otto von Bismarck, Pettenkofer did not.

Both emperor and chancellor much preferred Koch to Pettenkofer. Koch was not only a fellow Prussian, but they consistently found they could make much more political capital out of his discoveries. Campaigning under an anti-British ticket in the run-up to the Reichstag elections of 1883, Bismarck had relished exploiting Koch's claim that the Egyptian cholera epidemic might not have happened but for British commercial selfishness. Yet Koch achieved a status far beyond that of a regional or party political ornament; and this goes a long way towards explaining why he became Germany's public-health maestro.

In 1871, having crushed the French army, columns of Prussian-led German forces marched in triumph down the Champs Elysées. For the German peoples this devastating victory did much to expiate the sense of shame that had lingered on since their defeat at the hands of Napoleon in 1806. But in the aftermath of hostilities, as the French economy sagged under the terrible weight of war reparations, many Germans complained that another imbalance was in urgent need of correction. Despite having many of the best laboratories in the world, the achievements of German bacteriology were overshadowed by the remarkable successes of Louis Pasteur and his Parisian lab. If they were to catch up, the Germans needed a man of at least equal skill, genius, and flair.

Alas, Pettenkofer had little to offer in this respect. He was an extremely able epidemiologist with an international reputation. But he had little interest in the hunt for new microbes that was making Pasteur famous. Robert Koch, in contrast, was a bacteriologist. He was also supremely able, and his discovery of the tuberculosis bacillus in 1882 had given his nation a medical triumph about which it could feel proud. Two years later, his announcement that he'd also found the cholera germ confirmed him as the 'German Pasteur'. In reality, as we've seen, Koch's research on cholera was nowhere near as compelling as his study of tuberculosis. Patriotic Germans, however, weren't going to pay much attention to this. They wanted to believe that Koch had once again humbled the French in a field central to Gallic scientific pride. And so his ambiguous findings were elevated to the status of unambiguous truth, and the Reichstag declared that his discovery was 'regarded beyond the borders of our fatherland as a brilliant testimony to the persistence and thoroughness of German science'.

This grateful nation couldn't have chosen to elevate a more suitable champion. For Koch had an intense personal dislike of his French rival, no doubt hardened by the fact that in 1871 he'd spent several gruelling months as a surgeon amid the gore of a Prussian field hospital, suturing wounds inflicted by French rifles, artillery, and bayonets. But whatever Koch's motives, he was more than happy to feed his countrymen's appetite for scientific pugilism. Determined to uphold 'German pride', Koch accused Pasteur of 'contributing nothing new to science', and disparaged his contribution to a scientific congress as comprising no 'new facts' but lots of 'completely useless data'. Pasteur returned these insults in kind and with interest, either ignoring Koch's work or denouncing his results as facile. Their relations sank to a new low, however, when Koch effectively boycotted the jubilee celebrations for Pasteur's seventieth birthday. Koch once wrote that he especially treasured the medal given to him by Otto von Bismarck because he could 'wear it as if it were a military decoration'. He wasn't joking. Koch fought his battles bent over a microscope, but he still saw every major finding as another martial victory by Germany over France. Perhaps a little cruelly, we might modify Clausewitz's famous dictum on diplomacy by saying that to Koch, medical research was nothing more than the continuation of war by other means.

From this shamelessly patriotic point of view, Koch's cholera expedition was the greatest coup of his career. He'd not only discomfited the British; in one powerful stroke he'd also eclipsed the French. A team sent by Pasteur had found nothing of scientific value in Egypt and was hastily withdrawn after one of their number contracted the disease and died. In contrast, Koch and his colleagues hadn't rested until they'd identified what most Germans were more than happy to accept as the true cholera bacillus.

There is a time . . .

There's another sense in which the weaknesses of Koch's discoveries were offset by exquisite timing. Pettenkofer's belief that quarantine measures were unnecessary had strongly appealed to the German merchant classes during the 1860s and 1870s. Like their British counterparts, they objected to anything liable to hurt trade, and a ship trapped in a quarantined port could mean ruin for its owners. Pettenkofer's ideas also won over the authorities of several states who presented his opposition to quarantine as a mandate for taking hardly any action to improve public health. The city chiefs of Hamburg, for instance, loathed public expenditure. So, in the same spirit as they'd auctioned off the city's art collection, demolished its Gothic cathedral (pulverizing the statuary for use as building hardcore), and abolished church music as unjustified drains on the public purse, they adopted a crudely bastardized version of Pettenkofer's theory to justify doing the bare minimum to prevent epidemics. In contrast, Koch's work was of little use to them. His contagionist model afforded states little room for manoeuvre and implied that quarantine was an essential response to the outbreak of epidemic disease.

No matter how good his data, Koch's ideas would struggle to win through as long as the liberal, free-trade consensus remained intact. Luckily for him (and for many others), following the unification of Germany in 1871 a newly paternalistic political culture arose. The various states began to involve themselves much more frequently in their citizens' affairs and hundreds of new schools, museums, galleries, and hospitals were constructed throughout Germany. This new political climate strongly favoured the acceptance of Koch's contagionist germ theory. Keen to flex some political muscle, within a few years most state legislatures had passed quarantine laws. Pettenkofer's anti-quarantine stance, although based on a perfectly sound theory, now began to appear outdated and irresponsible.

But it wasn't only the states that became more active during the 1870s. A master of *Realpolitik*, Bismarck was usually careful to stay out of local affairs so as not to endanger the still fragile union he'd forged. At the same time, he was eager to find means of bringing the disparate states closer together. Public health, he recognized, was one way in which central authority could be exercised without offending the provinces too grievously. Disease hath no boundaries, Bismarck could argue, therefore it's essential for a central body to impose quarantine restrictions unhindered by state administrations. The Imperial Health Office, with Koch at its helm, served this vital political and medical function. Koch's theory had given Berlin a legitimate pretext for interfering at a regional level and binding the states together in a common endeavour. By emphasizing local conditions above contagion, Pettenkofer's theory offered the Berlin government no such political advantages. In consequence, the germ-soil theory became anathema, while Koch's contagionist germ theory enjoyed full backing.

Not much had happened during the 1880s to lessen the real scientific credibility of Pettenkofer's ideas. Koch's discovery of the comma-bacillus, coupled with his failure to account for Pettenkofer's epidemiological data, ought if anything to have strengthened his rival's position. This, however, would be to ignore politics. For in the short term, Koch's success relied on the fact that his ideas served the political interests of a proud and increasingly unified Germany in a way that Pettenkofer's theory did not.

Being fair to Pettenkofer

Superficially, Pettenkofer's self-experiment looks like the last throw of a man who knew he'd been bested and was willing to go to almost any lengths to buy himself a bit more time. But this is a judgement suffused with presentism. To become the astonishingly powerful set of ideas it is today, the germ theory of disease has had to undergo very considerable refinement over the last hundred or so years. In the 1880s it was dogged by problems that its pioneers simply could not resolve. Pettenkofer rightly drew attention to some of these deficiencies. In particular, he demonstrated that the germ alone is often not enough to induce illness, and that by concentrating almost entirely on the bacterial agents of disease Koch had told less than half the story. Far from being an act of puerile folly, Pettenkofer's self-experiment should be seen for what it really was: one of the critical moments leading to the birth of the modern science of immunology.

There is another sense in which Pettenkofer advanced the cause of medical science. As we've seen, Koch's single-minded obsession with germs inclined him to undervalue the importance of epidemiological data. The seductively simple formula that arose from his belief that germs alone cause illness led to a general neglect of a complex matrix of relevant factors, not least the drainage capacity of local soils. Even if he was ultimately wrong about how germs induce sickness, Pettenkofer's epidemiological studies identified genuine phenomena and began the difficult task of integrating laboratory findings with knowledge of the far messier outside world in which epidemics occur.

In the 1900s, bacteriology, immunology, and epidemiology all converged in the modern science of public health. The result has been a revolution in our understanding of the cause and spread of disease and the most effective means of preventing it. This immensely fruitful partnership is as much the legacy of Max von Pettenkofer as it is of Robert Koch.

PART TWO

EUREKA! REVISITED

EUREKA! REVISITED

[Sir Isaac Newton] was sitting alone, in a Garden, when some Apples falling from a Tree, led his Thoughts to the Subject of Gravity.

Benjamin Martin, *Biographica Philosophica* (1764).

The story of an apple falling in the vicinity of Isaac Newton while he sat in a quiet Lincolnshire garden in 1666, pondering planetary motions, is the first example of inspirational scientific discovery many of us encounter. For others, it's the case of Archimedes, in his bathtub, shouting 'Eureka!' when he realized that the volume of immersed bodies, including his own, could be worked out by measuring the amount of water they displaced. The fact that these stories are so memorable probably has something to do with the juxtaposition between a banal stimulus or setting, be it falling apples or an evening soak, and a sudden burst of world-changing creativity. Only a genius, one infers, could have derived a revolutionary insight in circumstances so humdrum. But for whatever reason, such stories continue to dominate the portrayal of science in the media, textbooks, and popular histories. Scientific progress, we're repeatedly informed, is largely the achievement of a handful of giants, those stand-alone heroes dubbed 'scientific Shakespeares' by the novelist and technocrat C. P. Snow.

The appeal of Eureka moments to hagiographers is obvious. They suggest that the discovery in question arose not from laborious plodding or patient inquiry, but from a realization so immediate and so profound that it demanded intelligence of the rarest kind. Scientists said to have experienced Eureka moments—Newton in the Lincolnshire garden, Galileo in Pisa Cathedral, James Watt when watching a boiling kettle, or Charles Darwin on the Galapagos Islands—are defined as geniuses in a tradition stretching back at least to the Renaissance and to the representation of the great artist as an individual blessed by God with a special gift of perception. This definition of genius was popularized by the Romantics, including William Wordsworth, who eulogized Newton as 'Voyaging throu' strange seas of Thought, alone'. Today, Newton continues to exemplify the idea of the lone discoverer gifted with uncommon intellectual powers.

The question is, how many of these iconic moments of discovery are actually borne out by historical evidence? In this section we re-examine three well-known episodes of supposed 'instant rationality'. First, the genesis of Isaac Newton's theory of light and colour in 1672; second, the surgeon James Lind's 1747 seafaring experiment in which he's said to have discovered that citrus fruits cure and prevent scurvy; and third, the hygiene reforms instituted in 1865 by the Hungarian obstetrician, Ignaz Semmelweis, in an attempt to reduce the appalling mortality rate from childbed fever in a Viennese maternity ward. The standard biographies of Newton, Lind, and Semmelweis almost invariably include a dramatic moment in which the truth dawned on their hero with astonishing clarity. And for all three, this breakthrough is said to have been embodied in a single experiment or demonstration of absolute transparency. The hardships they're supposed to have endured in trying to get their ideas across is then described in a manner perfectly consistent with the lone genius plotline. Both Newton and Semmelweis are said to have suffered severe psychological repercussions as a consequence of the opposition they faced. And the apparently shoddy treatment meted out to Lind and Semmelweis is believed to have cost thousands of lives.

But on closer inspection our stories serve only to question the traditional view of scientific progress as a heroic saga in which years of relative stasis were punctuated by short bursts of rapid advance. Each chapter reveals that experiments now celebrated as brilliantly decisive were in their own times nothing of the sort. Not until much later, with the accumulation of fresh data, new concepts, and far better instruments, did our heroes' experiments acquire real significance. Only then could they be rewritten as defining moments in the history of science on which only the foolish turned their backs.

Writers weaned on the romantic tropes of great-man history have not only tended to exaggerate the clarity of our subjects' experiments. With respect to Semmelweis and Lind, many have also overstated the similarities between their ideas and modern orthodoxies. Both men made assumptions so remote from the way we think today that it has taken scholars considerable effort to piece together their world-views and reassemble the logical structures of their arguments. This kind of painstaking reconstruction is vital if we're to understand the genesis of new ideas and the conduct of scientific debate. Blundering in with a ready-made template comprising a Eureka moment, a decisive experiment, and an ignorant opposition really won't do. In fact, it's the historiographical equivalent of the way generations of archaeologists fitted in virtually every scrap of archaeological data found in Palestine to Old Testament stories.

Before moving on to our three case studies, we might briefly examine how these themes relate to the classic moment in the history of science with which we began: Isaac Newton and the falling apple. The first thing to note here is that even if the apple story describes a real event, which many historians doubt, it couldn't possibly have marked the moment at which Newton arrived at his theory of universal gravitation. We can be quite sure of this because in 1666 he was still almost fifteen years away from getting there. Having begun thinking about planetary motions early in the 1660s, it wasn't until about 1680 that Newton's theory finally came together. It had required in all nearly twenty years of gestation. While this rate of progress was anything but pedestrian, there had been no blinding flash of insight during which the theory appeared fully formed in Newton's mind. Presumably he did experience a certain *frisson* when all the pieces first slotted into place, but by then he'd been accumulating the key components for many years. Nor, during this time, had he been working in a vacuum.

Newton was well versed in the works of Descartes and other leading thinkers of the Scientific Revolution by the time he'd finished his undergraduate studies. He also gained invaluable mathematical concepts from the works of the astronomer Johannes Kepler. In addition, we know that during 1679 Newton began a short-lived correspondence with the Royal Society curator of experiments, Robert Hooke, on the very subject of planetary motions. It was in the immediate aftermath of this exchange of letters that Newton finally settled on the theory he would publish in his *Principia Mathematica* of 1687. It's highly suggestive that before this correspondence, Newton doesn't appear to have been thinking in terms of bodies exerting attractive forces upon one another from a distance. Indeed, he was probably striving for very much the reverse, assuming at one stage that the moon maintained its orbit due to unseen mechanical restraints rather than some kind of immaterial force. Hooke, however, in a letter to Newton of 1679, referred explicitly to the notion that 'an attractive motion towards the centrall body' was responsible for planetary orbits. Shortly after, Newton formulated his theory of universal gravitation. To most historians, Newton unquestionably *was* the chief architect of the theory of universal gravitation, not least because he could place it on a mathematical foundation way beyond the competence of Robert Hooke. But many have come to the view that Hooke deserves to be seen, at the very least, as a midwife to its discovery.

For all of Newton's individual brilliance, the path from initial insight to final theory was long and arduous. It also required a familiarity with existing theories and a timely push in the right direction. A genius Newton might have been, a lone searcher after the truth he was not. Nor, it should be added, could his theory be judged obviously correct in the late 1600s or even in the early 1700s. As we saw in the chapters on Glanvill and Spallanzani, several eminent natural philosophers deplored his invocation of immaterial laws. Even those who found these palatable were often unwilling to commit themselves until it was proven that his hypothetical laws fitted the observed motions and trajectories of the planets. This crucial process of validation took several decades to complete. So, far from Newton overseeing the birth of a new kind of physics, he went to the grave before his ideas could be declared proven beyond reasonable doubt.

Newton first told the apple story almost three decades after it was supposed to have taken place. Given that his theory had only come together a few years before, it's quite likely that he knew the story to be at best a half-truth. Several historians have therefore wondered why he told it. For one thing, we know that in Newton's time the apple had tremendous symbolic importance. If he wished to fashion a poetic context for his discovery, linking it to the biblical tree of knowledge from which Eve plucked an apple was a natural choice. But why invent a Eureka moment to begin with? One suggestion is that it was fashioned in direct response to Hooke's claims to a share of his spoils.

When he first saw a copy of Newton's *Principia* Hooke angrily declared that his own vital contribution, contained in his 1679 letter, ought to have been properly acknowledged. Consumed with anger at what he saw as gross effrontery, Newton thereafter nursed the bitterest hatred for Hooke, going so far as to withdraw from all involvement in the Royal Society until after Hooke's death in 1703. Some historians are convinced that Newton despised Hooke with such passion that he simply couldn't tolerate his gaining any credit for the theory of universal gravitation. He therefore devised the apple story. This backdated the discovery by fifteen years and so put Hooke out of the picture. If this is true, and it's a perfectly plausible theory, then the strategy was a resounding success.

In a few paragraphs we've seen that there are lots of problems with the received account of Newton discovering the laws of gravity. But the true story, for all its added complexity, is surely no less engaging than the legend. We forfeit the concentrated drama of the Eureka moment, but we gain a lead character many times more interesting than the bloodless hero of popular mythology. We also acquire a truer sense of the scientific endeavour as a multi-participant event involving many sharp minds, not just the occasional extraordinary genius.

The Antechapel where the Statue stood
Of Newton with his prism, and silent face,
The marble index of a Mind for ever
Voyaging thro' strange seas of Thought, alone.
                    William Wordsworth,
                    The Prelude (Book III, 1850).

It was an age of experiments. When Isaac Newton was a boy, the king had been beheaded and replaced by a republic presided over by a gifted soldier born into the middling orders. With the restoration of the monarchy in 1660, there then arose an ultimately far more fruitful kind of experiment. For millennia natural philosophers had condemned those who indulged in scientific experimentation. Castigated as mere 'empirics', their crude manipulations were contrasted unfavourably with the brilliant webs of all-encompassing theory spun by true philosophers. Philosophy, the intellectual elites declared, is all about thinking not doing. But the newly emerged scientific ethos of the seventeenth century changed all that. The untested and untestable theories of conventional natural philosophy were now seen to yield little more than barren intellectual feuds. England after the Civil War was in no mood for this kind of internecine conflict in either politics or science. And the solution devised by her leading natural philosophers was to have far-reaching implications: the middle years of the century saw them adopt what has become the touchstone of modern science, the experimental method.

As we saw in the chapter on Joseph Glanvill, this new approach to science was embodied in the protocols of London's Royal Society. Members of this society were still allowed to theorize, but no theory was to be upheld if it couldn't be proven experimentally before a gathering of competent witnesses. At the Society's meetings, natural philosophers

LEFT: *Study for a 1725 portrait of Isaac Newton.*

congregated to observe new experiments. Before leaving they were expected to reach a clear consensus as to what they'd seen and what, if anything, it proved. The impact of these demonstrations spread well beyond the walls of the Royal Society. Experimenters published descriptions of their methods and apparatus in the society's *Philosophical Transactions*, and this enabled others to repeat key experiments before other scientific societies elsewhere in Europe. Science done in this way, enthused the Royal Society's first historian, Thomas Sprat, is free from the subjectivity of traditional philosophy. It allows one to understand the world with nothing short of 'Mathematical plainness'.

It was in just this spirit that in 1672 the young Isaac Newton wrote to the Royal Society announcing what he believed to be an 'experimentum crucis', literally a crucial experiment. It was supposed to be a demonstration of such breathtaking clarity that only the ignorant or bloody-minded could fail to arrive at the proper conclusion. Newton's experiment was intended to show that sunlight comprises a bundle of distinct rays, each distinguished by a different colour and angle of refraction. Three hundred years later, this explanation for rainbows, the fans of coloured light that emerge from prisms, and the spectral hues visible in oily puddles is not too difficult to grasp. But in the 1670s, when it was first unveiled, Newton's theory of light and colour was little short of revolutionary.

Most of Newton's fellow natural philosophers believed that the colour of light, be it white, red, blue, or violet, was determined by the speed at which it travelled through the atmospheric æther. The fastest pulses were usually said to be white and each position on the spectrum, from red to violet, was associated with an incrementally lower velocity, finally ending with pure darkness. Rather like their contemporary alchemists who sought to convert one element into another, natural philosophers assumed that coloured rays of light were transmutable. To change blue into red, white into yellow, or orange into violet, they reckoned that one simply had to find a way to quicken or retard the speed at which the pulses moved through the æther. One way of doing this, they surmised, was to pass them through a prism. This was the kind of thinking that Newton's experiment was designed to overturn. And according to most, it did just that.

Newton's *experimentum crucis* is often seen as experimental science at its devastating best. Thus, for some of the Romantic poets, his 'unweaving of the rainbow' symbolized the awesome power of experiment to demystify nature and to unlock her most subtle mysteries. John Keats, for instance, is said to have bemoaned Newton's influence 'because he destroyed the poetry of the rainbow by reducing it to a prism'. There has been a tendency to overemphasize the Romantics' disapproval of Newton's efforts. But one thing is very clear: Newton's hard, uncompromising science seems utterly at variance with the subjective nuances of the poetical imagination.

As this chapter seeks to show, however, Isaac Newton's famous experiment of 1672 was really far less conclusive than he and his allies claimed. Cutting-edge science is usually a messy, difficult, and uncertain business, and in the half-light in which it takes place the experimental evidence is rarely easy to interpret. Many years later, when closure is reached and the theory is finally deemed obviously true, we forget how uncertain things once were. Those in the right are then hailed as heroes and their erstwhile rivals damned as bloody-minded egoists. Yet while feuds in science might be intensified by jealousy, ambition, and ill-will, they are usually sustained by genuine concerns about method and theory arising from the fact that leading-edge experiments are nearly always open to multiple interpretations. This chapter draws on the research of the scholars Harry Collins, Trevor Pinch, and, in particular, Cambridge University's Simon Schaffer, to show that Newton's *experimentum crucis* was not only indecisive, but that it had emerged victorious by the time of his death, in part due to the despotic grip he exercised over the Royal Society.

The experimentum crucis

Aged 24 and entirely unknown to the wider scientific community, Isaac Newton had just completed his bachelor's degree when, in 1664, he started to question conventional theories of light and colour. He began by passing a beam of light through a prism and projecting its spectrum onto the rear wall of a large room in Trinity College, Cambridge. Instead of seeing a circle of colour, as he expected, Newton saw an oblong of light in which 'ye rays wch make blew are refracted more y^n ye rays wch make red.'

It was a portentous finding. As we've seen, most natural philosophers assumed that spectral hues emerged from prisms because the prismatic glass altered the speed at which the pulses of light travelled. But this explanation for the appearance of coloured light gave absolutely no reason to expect the pulses to be refracted at different angles. The full importance of Newton's observation, however, only dawned on him two years later when, in his suite of college rooms, he was trying to devise ways of grinding elliptical lenses. Abandoning the 'aforesaid Glassworks', he later recalled, 'I procured me a Triangular glass-Prisme, to try therewith the celebrated Phaenomena of colours.'

Over the following weeks he devised, refined, and replicated dozens of experiments, finally arriving at a startling conclusion: 'Light it self is a Heterogeneous mixture of differently refrangible rays.' In his famous 1672 letter to the Secretary of the Royal Society, Newton related how he'd seized upon the remarkable idea that white light comprises a blend of coloured

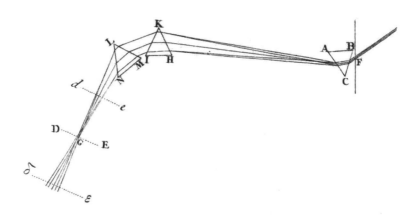

An idealized image of the experimentum crucis *from the third edition of Isaac Newton's* Opticks, *published in 1721. Light enters on the right-hand side ('F'), and passes through a prism which refracts the beam's constituent rays. These uncompounded rays continue through further prisms and remain the same colour throughout. The various rays are then focused, giving rise to white light, before being refracted again.*

rays. And although he spoke of his efforts as 'poore & solitary' he was thoroughly convinced that he'd got it right and that his theory was of tremendous importance.

Nevertheless, Newton recognized that 2000 years of misconception are not easily overturned, so he set about designing an experiment that he believed would deprive his peers of any prospect of disagreement. He began by boring a hole in his 'window shuts, to let ... a convenient quantity of the Sun light' into his 'darkened chamber'. A narrow shaft of light now passed first through the hole in his shutters and then through a prism suspended close to the window. The light's constituent coloured rays were refracted into a broad spectrum of colour by this prism and Newton was able to separate out a single, 'uncompounded', coloured ray by placing in front of the first prism a solid board with a tiny hole drilled through it. This lone coloured ray travelled through the hole in the board and then onto a carefully positioned second prism. As it emerged from the other side, the ray's hue remained unchanged and its angle of refraction consistent. No matter how many additional prisms Newton passed his isolated ray through, it retained these identical properties. As he explained, 'when any one sort of Rays hath been well parted from those of other kinds, it hath afterwards obstinately retained its colour, notwithstanding my utmost endeavours to change it.'

From this experiment Newton drew two daring conclusions: that the spectrum is made up of primary or 'primitive' colours that are fixed and immutable, and that when these primitive rays are combined the result is what we know as natural light. Newton's 1672 letter was brought to public attention later that year. Many philosophers were stupefied at the brilliance of this young natural philosopher and effusive in their admiration. Newton, one declared, is 'our happy wonder of ingenuity and best broacher of new light'.

A papist plot?

Newton's 1672 letter to the Royal Society brought to a close a long period of intellectual isolation. Fearful of controversy and unwilling to engage in open debate, he hoped devoutly that his fellow natural philosophers would now perceive the logic of his explanation and glean from the *experimentum crucis* the message it was designed to convey. And, to a large extent, this is exactly what happened after it was performed in 1675 before the assembled ranks of English natural philosophers. Yet, from the outset, there were dissenting voices.

Throughout the 1670s Newton was subjected to unrelenting criticism from a group of English Jesuits working in a seminary in the town of Liège, then part of the Holy Roman Empire. Led by Francis Line, a mathematics professor, John Gascoines, his student, and the theology professor Anthony Lucas, they wrote to Newton explaining that they had performed dozens of prism experiments but had been unable to replicate his findings. Their persistent criticisms quickly drove Newton into an almost incandescent fury. Having replied courteously to Line's first letter, in the second he demanded that his correspondent cease 'slurring himself in print'. To Father Lucas he was more brutal. Your criticisms, he raged, are 'too weak to answer'. After another letter, this time from Lucas, Newton refused to accept any communications emanating from Liège.

To generations of historians, Newton's viciousness towards these Jesuits has been seen as an acceptable manner for geniuses to treat stubborn and officious amateurs. And the effrontery of the Liège Jesuits in attacking the great man has brought much posthumous contempt upon them. One historian has disparaged Line as a 'narrow-minded pedagogue'. A Newton biographer scathingly remarked on 'a Belgian named Linus, who was a stupid, ignorant and narrow-minded man'. And in 1980 yet another biographer referred caustically to the 'ignorant Papists' of Liège. Indeed, the Catholicism of this group has often been implicated in their hostility to Newton. Becoming increasingly paranoid during the summer of 1678, Newton himself thought their critiques part of a dastardly papist plot to destroy his reputation. Yet there is no serious reason for thinking that either religion or stupidity had anything to do with this controversy.

After all, since its foundation in 1540 the Jesuit order had been renowned for its reverence for learning. To this order of pious educators, the discovery of natural laws served to enhance faith by revealing the sublime coherence of the divine plan. And the Liège seminary took its intellectual function with the seriousness befitting what was in effect practical theology. 'Sedulous in opening up the secrets of nature through experiments,' boasted a 1665 manifesto, 'there is nothing . . . which has been ingeniously discovered by the Royal Society . . . which has not been taken up and adorned by members of our college.'

Given that the Liège Jesuits were dedicated and competent men of science, we need to look more closely at their criticisms. Father Line had begun by trying to work out if prisms really do project oblong spectrums. He approached the question with impressive rigour. 'We think it probable he hath tried his experiment thrice for Mr. Newton's once,' reported one well-placed observer. In his 1672 letter Newton had declared that this oblong was always five times longer than its breadth. But on each occasion Line tried the experiment, the shape of the light on his wall was closer to the circle predicted by traditional philosophy than the oblong with a constant ratio Newton claimed to have found. Line speculated that perhaps Newton had erred by doing the experiment on one of England's not so rare cloudy days, with the result that the clouds themselves, rather than the prism, had dispersed the sun's rays. Line was wrong, but it was a fair suggestion.

Soon after Newton broke off contact Line died, according to Gascoines, of the 'epidemical catarrh, which has raged through so many countries'. His place in the front ranks was taken by Father Lucas, and Lucas too had immense difficulty in reproducing Newton's results. In reply, Newton implored him to 'try only the experimentum crucis. For it is not number of Experiments, but weight to be regarded.' Lucas dutifully obeyed these instructions and undertook an arduous course of experiments. Still, however, his results did not tally with Newton's. Every time Lucas attempted to replicate Newton's work he failed to produce rays that stayed the same hue after the second refraction. Despite setting up his equipment 'exactly according to Mr. Newton's directions' he found 'as a result of many trialls' that violet rays formed at the first refraction displayed a 'considerable quantity of red ones' after they'd passed through the second prism.

This looked very bad indeed for Newton. The Jesuits had seemingly mastered the technique of isolating single coloured rays using the first prism, but they'd confounded his predictions by showing the beam changing colour at subsequent refractions. Clearly, Newton thundered, something was wrong with their prisms. But this was not the only possible explanation. As the Jesuits recognized, the appearance of additional colours at the second refraction argued against there being such things as primitive colours. On this evidence, it seemed much more likely that the second prism had somehow accelerated the speed of the violet pulse and thereby made it appear red. This was an interpretation with which at least one leading English philosopher had considerable sympathy.

By no means conclusive

Robert Hooke and Newton were bound to fall out. In his 1665 book, *Micrographia*, Hooke probably thought he'd said pretty much the last words on light and colour. Then Newton made his sudden appearance from the relative intellectual backwater of Cambridge University. For the remainder of his life, Hooke accused Newton of plagiarism whenever he agreed with him and put up fierce resistance wherever he didn't. But this was never simply a question of bad sportsmanship on Hooke's part. Although he tends to be cast as Newton's jealous rival, he was highly accomplished in his own right and with some justice felt wronged by Newton. He was also an acute critic. And in the case of the *experimentum crucis*, Hooke posed some genuinely awkward questions.

In particular, he asked how Newton could possibly demonstrate that *before* the first refraction sunlight really contains a bundle of primitive rays. Newton believed that he was using prisms to separate light into its constituent parts. But it remained possible that the prism itself caused the appearance of spectral light by modifying rather than disaggregating the natural light beams. On this basis, in 1675 Hooke presented his own theory according to which differently coloured rays of light are produced by vibrations of different degrees of intensity in the atmospheric æther. The action of passing through prisms, he argued, generates such turbulence that many different colours emerge and are refracted, as Newton had shown, at different angles. 'The same phaenomenon will be salved [i.e. explained] by my hypothesis as by [Newton's],' Hooke confidently remarked.

Hooke was prepared to concede that coloured light beams produced by prisms can maintain the same properties in subsequent refractions (he had himself managed to replicate the *experimentum crucis*) but he added that this was only one of many ways to make coloured light. In 1666 he explained how he'd projected a beam of prismatic blue light onto a piece of red cloth. This gave him the impression of looking at purple light and convinced him that here was another way in which coloured beams could be generated. Knowing nothing about how different wavelengths of light can travel together in the same beam, his was a perfectly reasonable inference to draw.

Locked into a feud with Newton, Hooke was also delighted to hear of the evidence compiled by the Liège Jesuits of new colours appearing at the second refraction. He was no less pleased when the Frenchman, Edmé Mariotte, an experienced investigator of optical phenomena, stirred up more trouble for Newton. In 1676 Mariotte set up the apparatus for the *experimentum crucis* in his Parisian laboratory. He made every effort to separate a single ray from the first refraction by passing one colour through a tiny slit in a piece of white card set thirty yards away from his first prism. But Mariotte observed that a ray emerging from the first refraction that seemed pure violet exhibited red and yellow tinges after it had travelled through the second prism. Once more, it seemed that Newton had been experimentally disproved. As will be seen shortly, the Frenchman's odd results were probably caused by minute bubbles in his first prism. But for Mariotte there was another, much more compelling explanation: the idea that natural light is compounded of differently coloured rays is wrong.

Not to be left out, the Italians eventually entered the arena. And in science as in politics, more than one voice emerged. In Bologna, the experimentum crucis was successfully replicated. In Venice, however, things went far less smoothly. After reading Newton's *Opticks* in 1719, the natural philosopher, Giovanni Rizzetti, took care 'to repeat all [Newton's] experiments'. He then declared 'some false and all the rest equivocal and by no means conclusive'. The centre-piece of Rizzetti's attack was his own attempt at the *experimentum crucis,* in which rays that seemed pure yellow after the first refraction had become red, green, and indigo after the second. It was an extraordinary result and it gave Rizzetti the confidence to go on challenging the Newtonians for several years.

Indeed, historians have shown that even four or five decades after

Newton's 1672 letter no absolute consensus as to the nature of light had yet been established. Texts on optics still respectfully cited the work of Mariotte and Rizzetti. Moreover, Newton's supporters held sway only where their authority could be most firmly stamped: the Royal Society and European scientific institutions whose members had witnessed the experiment in London or had been able to reproduce it for themselves. Despite this, most historians and Newton biographers have reserved only contempt for his critics. For them, this debate was a futile sideshow that wasted too many of Newton's productive years, caused him inexcusable annoyance and emotional upset, and could have been avoided were it not for Line, Lucas, Hooke, Mariotte, and Rizzetti's twisted personalities. With relish, several commentators have quoted the colourful judgement of Newton's champion, the French Protestant refugee, John Théophile Desaguliers, that Rizzetti's poor-quality prisms had 'rendered him ridiculous for ever'.

Yet this assessment could only have been made with the dubious benefit of hindsight. The biographers who have disparaged Newton's opponents all assume that his *experimentum crucis* was indeed decisive to anyone whose judgement was unclouded by jealousy or sectarianism. But, as the historian Simon Schaffer has shown, the kind of clarity that his experiment has for us 300 years later was not so apparent at the time. Then, there was much to cloud the issue.

Following directions isn't easy

There is often a dramatic difference between what a scientist does when performing an experiment and the account of it that appears in print. This is partly because the methodology sections of journals would be excessively long if they described absolutely everything. But it's largely due to the fact that, during the course of an experiment, all scientists do things that are so habitual that they don't even think of mentioning them. In most fields of scientific enquiry, this doesn't present a major problem. Routine procedures, chemicals, receptacles, drugs, even laboratory animals are now so standardized that replicating an experiment in another laboratory is usually fairly straightforward.

Yet this is *not* the case with cutting-edge science. Here, anything the scientist holds back can make it very hard for other laboratories to repeat the original experiment. Indeed, it's sometimes necessary for personnel from one laboratory to go to another to provide directions on the spot. Scholars refer to the missing ingredients in many scientific accounts as 'tacit knowledge' and Newton's account of his *experimentum crucis* illustrates just how much confusion incomplete methodologies can introduce.

Newton's 1672 letters to the Royal Society gave wholly inadequate directions for replicating most of his experiments. For the majority of the tests he mentioned, the distances specified for the gaps between prism and window, board and prism, and prism and wall were only vaguely expressed. And, despite the enormous variation in commercially available prisms, Newton confined his instructions on their purchase to the unhelpful remark that they should be 'clear and colourless'. Only the most determined and patient natural philosophers managed to produce primitive rays on the basis of this spartan commentary, and Newton had to concede by the end of the year, 'I am apt to believe that some of the experiments may seem obscure by reason of the brevity wherewith I writ them.' Over the following months he elaborated on his original descriptions. But two years later the *experimentum crucis* remained stubbornly problematic. Once more Newton admitted that it was 'not yet perfect in all respects'. In some of his own experiments, he went so far as to confess, positioning prisms in certain ways caused red rays to acquire yellow colouring at the second refraction. As correspondents continued to complain of their difficulties, he began to see his error more clearly. Just prior to breaking off communications, Newton gave the Jesuit Anthony Lucas more details about the quality and angles of the prisms, the amount of daylight required for the experiment to work, and then implored him to abandon the slightly concave prism he had been using and instead use one with convex sides. He also explained that most prisms are unsuitable because they contain so many 'bubbles', 'veins', and 'impurities' of colour.

Even this flurry of additional advice proved insufficient. And it was partly because of Newton's realization that he and his opponents were discussing essentially different experiments that Newton published his *Opticks* in 1704. In this book Newton finally gave an appropriately thorough account of the methods and instruments required to perform the *experimentum crucis*. He also added entirely new procedures, such as the placing of a lens before the first prism in order to focus the sunlight. In addition, Newton did his best to give clear specifications for selecting prisms. The polish had to be 'elaborate' and not 'wrought with putty', which caused 'little convex . . . Risings like waves'. The edges of the prisms had to be covered with black paper. And, admitting that 'it's difficult to get Glass prisms fit for this Purpose,' Newton noted that he'd often had to fall back on 'broken Looking-Glasses' filled with rainwater, and lead salt to increase the degree of refraction. Perhaps most remarkably, he flatly contradicted an earlier prescription: prisms with 'truly plane' sides were to be used, not the convex kinds he'd recommended to the Jesuits. Reflecting on these meticulous directions, his ally Desaguliers noted that a method for reliably separating primitive rays 'was not published before Sir Is. Newton's *Opticks* came abroad'. Several decades of refinement had transformed the *experimentum crucis* into a much more reliable, but also a rather different exercise than the one he'd described in his letter of 1672.

So for over thirty years, Newton's troubles with light were in large part self-inflicted. The poverty of his first descriptions and, almost certainly, defects in his original procedure ensured that some perfectly able would-be replicators failed to find what he'd seen. But even after the tacit knowledge required to perform the experiment and the refinements he'd developed had at last been put into print, Newton's troubles were still not at an end. His experiments with prisms then came to highlight a chronic difficulty with the experimental method, which made his feuds with Mariotte, Lucas, and the rest both inevitable and very hard to resolve.

Experimenter's regress

Scientific theories, according to the canonical scientific method, are supposed to be formulated on the basis of experimental results alone. It's supposedly mandatory for practitioners to abandon their theories if they don't accord with the outcome of experiments. As Newton's peers in the Royal Society would have chorused, experiments rule and theory must slavishly follow. But things in science aren't always so uncomplicated.

Since newly discovered phenomena require new methods of investigation, scientists working at the outer margins of knowledge have to develop their experimental apparatus as they go along. The problem here is that novel experimental protocols involve high degrees of error and throw out lots of rogue results that have nothing to do with nature and everything to do with faults in the experimental set-up. Dust, for instance, might interfere with machinery and produce strange results, or metal pins used in setting up pieces of apparatus might upset magnetic fields. Yet until the unknown contaminant has been identified and removed, it's not possible to decide for sure whether an unusual result is reliable and important or an error and therefore meaningless. Clearly, one can't automatically declare an odd data-set to be of no value, because scientific discovery feeds on anomalous, unexpected outcomes. Nor can one say that rogue results must be of genuine value, since all scientists know that cutting-edge experiments tend to have teething troubles.

Trevor Pinch and Harry Collins have called this phenomenon 'experimenter's regress' and they've described several striking examples in their book, *The Golem*. Of course, the only way around such difficulties is to study the supposed phenomena from as many angles as possible with such a variety of experimental tools and techniques that one can eventually say with confidence whether it's real or merely an artefact of the equipment used. But this takes time. And in the weeks, months, or years before a consensus is achieved, disputes with a highly distinctive pattern are likely to emerge.

Team A says that their unexpected result should be taken seriously. Team B fails to replicate it and declares it to be bogus. Team A replies that B obviously set up the novel apparatus incorrectly. Team B retorts that Team A is guilty of special pleading. In A's view, B will only know that they've assembled their experiment correctly when they have reproduced A's result. But from B's perspective this might mean building any original experimental flaws into their own set-up. So, for Team B, the argument that their experimental equipment can only be judged valid when they manage to confirm A's findings seems to be an absurd example of circular reasoning. Yet both A and B are making potentially valid points. Team A might be on to something important. On the other hand, they may have made an unfortunate error. For a time there really is no way of telling whether the experimental results should be accepted or ignored and the outcome of the experiment remains the only way of telling if the two teams are really following the same procedures. This was precisely the logical bind in which Newton found himself.

An unhappy choice of prisms

The basic problem for Isaac Newton was that in 1672 the prism simply was not a standardized scientific instrument. Symptomatic of its lowly status, Newton seems to have procured his own in one of the few places in which they were easily available: country fairs. Often called 'Fools Paradises', prisms were favourites of fairground magic men; the ability to produce an array of spectral light using an angled block of glass could hardly fail to impress the uneducated with the powers of the performer. But even the educated were susceptible to the prism's charms: in some nurseries they were used as toys and in certain dining-rooms they added impressive lustre to chandeliers.

Newton was therefore seeking to make an article of decoration and popular amusement the core of his *experimentum crucis*. For a long time prisms had been an object of interest to natural philosophers, but no attempt had been made to study how different prism geometries or impurities might affect the behaviour of light passing through them. Furthermore, manufacturing prisms of standard sizes, shapes, and refractive powers had not even been suggested. Newton's own notes and letters reveal that there were dozens of different sorts of prism from which to choose. Some were smoother and more polished than others, most were coloured, and others contained trace elements of minerals such as lead. In consequence, there was no agreement as to what a good prism looked like or how light passing through it might be expected to behave.

It's now apparent that the prism used by Francis Line in Liège had a far lower dispersive or refractive power than any of those used by Newton. As a result, when he tried to project an oblong of light onto a wall he could only get something that looked like a squashed circle. Newton hurt his cause by coldly rebuffing Line's suggestion that the difference in results might have something to do with the 'specific nature of the glass'. But almost a century later a Fellow of the Royal Society demonstrated that Line had been correct in his suspicion. How Newton failed to realize at the time that the refractive power of different prisms varies is hard to understand. After all, he sometimes used prisms filled with water to which he added lead salt in order to increase light dispersion. Clearly, however, during the 1670s Line was perfectly entitled to trouble Newton with his doubts. Since there was no way of telling who had the most 'appropriate' prism, nobody could say whose result should be accepted and whose ignored.

An even more serious problem was that the quality of prismatic glass varied enormously. Veins and air bubbles, microscopic cracks, and barely perceptible coloured tints ensured that no two prisms were identical. Of course, none of this would have mattered if all but the most obviously poor-quality prisms gave similar results. But even very slight imperfections can significantly alter the pattern of light that departs from the prism. Knowing this, we can begin to understand why so many natural philosophers failed to replicate Newton's findings despite doing their utmost to follow his protocol.

In the case of Lucas' 1676 experiment, we can be fairly sure what went wrong. Within the sunlight passing through his first prism, a few of the bundles of waves making up the band of red light struck a microscopic bubble or vein in the prismatic glass. This produced a second refraction, and these fragments of red light happened to be deflected into the path of the band of blue light and became mixed up with it. Because there were only a few waves of red contaminating the blue light, these were imperceptible when the isolated blue beam passed between the first and second prism. Yet at the second prism another refraction took place, and the red rays then split off at a different refractive angle. Now that they could be seen, their presence falsely implied that the second prism had created them. And, given the limits of seventeenth-century knowledge, this was an entirely logical deduction for Lucas to make. Similar defects in Rizzetti's and Mariotte's prisms almost certainly explain their findings as well.

The filter of the mind

Newton himself eventually came to believe that impure prisms were the greatest obstacle to the resolution of the controversy. He implored Continental philosophers to abandon prisms manufactured by the highly regarded Venetian glass-blowers and use instead those he declared to be of more 'pure a Cristall', made in England. Newton also worked hard to entice the philosophical elites of Europe to the Royal Society, where the *experimentum crucis* could be reliably executed. But there were sound reasons for rejecting his suggestions. Looked at without the luxury of modern understanding, far from strictly observing the experimental method we can now see that Isaac Newton was actually inverting it. Had he developed a theory of light on the basis of his raw experimental data alone, it would not have looked like the one he presented in his 1672 letter.

Newton, just like Lucas, Mariotte, and Rizzetti, sometimes saw additional hues appearing in isolated rays of coloured light after the second refraction. He even admitted in his *Opticks* that it had taken him years to refine his experiment and its apparatus to the point at which he could ensure the emergence of only primitive rays from the second prism. Much of the raw data, in other words, simply did not support Newton's theory of light and colour. Lucas, Mariotte, Rizzetti, and Hooke realized that when Newton examined his prism data, they were filtered through another lens located somewhere in his own mind. Unswervingly committed to his theory, Newton accepted only the results that were consistent with it and rejected all others. If the experiment 'failed', Newton reasoned that the offending prisms must be flawed. But as we've seen, there were no objective criteria for assessing the quality of prisms to which he could appeal. So, with the circularity often considered the reverse of good science, Newton was judging his experimental output according to a theory that his experiment was supposed to be evaluating.

Conveniently unaware of this paradox, Newton and his allies proceeded to sift through an array of prisms until they found ones that gave the 'right' result. These were then taken as the standard against which all other prisms should be judged. Seen from this point of view, one can appreciate why Lucas, Mariotte, and Rizzetti bridled at Newton's insistence that they too go through their collection of prisms until they found one that gave them primitive rays at both refractions. Even though he had no way of proving that anything was actually wrong with most prisms, Newton was demanding that they abandon any that did not corroborate his theory, itself based upon a highly selective data-set.

So, just like our scientific Team A, Newton was telling his opponents that they would know when they had the proper apparatus because they'd be able to confirm his theory. Lucas, Mariotte, and Rizzetti, as is only to be expected of Team B, recognized Newton's overt bias and denied that his data had any special status. Instead, they happily published contrary results in support of the traditional theory of colour as modified light.

Newton wasn't behaving wholly irrationally. Where the prism and the phenomena of light were both badly understood, it made no more sense to take nearly all of the evidence seriously (as Newton's opponents did) than to deny the validity of all but a part of it (as did Newton himself). For example, those who adopted Hooke's rival theory of light and colour still had to explain why sometimes, using certain prisms, the *experimentum crucis* actually worked. Just like Newton, their standard response was to impugn the prisms used.

The manufacture of prisms gradually improved during the eighteenth century to the point at which most could be declared free from imperfections. At last it became possible to perform the *experimentum crucis* without difficulty, and Newton's theory then acquired a new plausibility. But the appearance of quality prisms did not coincide with the general adoption of his concept of light and colour. As Simon Schaffer has pointed out, except for a few pockets of resistance in Italy, Germany, and France, Newton had largely clinched the debate by the second decade of the eighteenth century, long before the standardization of the prism. This raises an intriguing question: how did he manage to persuade his fellow natural philosophers that his prisms, and not those used by his rivals, gave the most reliable results? The answer lies in a combination of Newton's genius, his autocratic personality, and the elevated status of the Royal Society.

El Presidente

Newton has justly been described as the first public scientist. In an age that had begun to set a high value on science, his extraordinary achievements in mathematics, optics, dynamics, and astronomy thrilled his countrymen and bathed his fellow philosophers in reflected glory. Within just a few weeks of his demise the hagiographers got to work. Realizing that they could exploit Newton's fame to promote a positive image of science, natural philosophers carefully refashioned his reputation. Newton became a model of both scientific genius and human decency, a man whose life epitomized the natural philosopher as a selfless and humble searcher after the truth. It was an image William Cowper happily embraced:

> Patient of contradiction as a child,
> Affable, humble, diffident, and mild,
> Such was Sir Isaac.

Cowper perhaps assumed that anyone as wondrously talented as Newton could afford to be gracious and munificent. But in this case at least he was

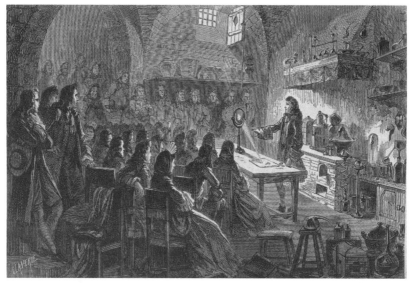

A person identified as Sir Isaac Newton showing an optical experiment to an audience in his laboratory. Wood engraving by Martin after C. Laverie.

very wrong. Diffident Newton could be, patient of contradiction and mild-tempered he certainly was not. As numerous contemporaries recorded, Newton had a habit of forming occasional intense relationships with other men that nearly always ended in a sense of betrayal on both sides. In the words of one erstwhile friend, William Whiston, he was 'the most fearful, cautious, and suspicious temper that I ever knew'. Newton was also notoriously stingy in recognizing the intellectual contributions to his own work made by other philosophers, and he fought a series of vicious priority disputes with unsavoury zeal.

The advantage Newton had was his control of the Royal Society, over which he presided between 1703 and 1727. There he operated with such dictatorial severity that only the brave ever gainsaid him. To those he upset by acting the martinet, Newton became the subject of intense dislike. But irrespective of how they felt about it, the Royal Society under his presidency became less a democratic forum for scientific experimentation than an elevated soapbox for Isaac Newton. And as his prestige as a philosopher rose alongside his position in the Society, Newton found that he had considerable power to steer the opinions of its members in his favour. It was partly this authority that enabled him to marginalize his rivals in the optics debate. President Newton took up many hours of the Society's time in performing finely engineered prism experiments before dozens of domestic and Continental philosophers. Such was the status of the Royal Society and the visual power of his experiments that nearly everyone went home convinced that Newton was right. He then capitalized on this by publishing the results in the form of official texts carrying the assent of virtually all the leading men of British and Continental science. No longer the hobby-horse of one man, his theory was validated by large numbers of competent and supposedly disinterested witnesses.

Aside from impressing his peers with his intellectual virtuosity, Newton maintained his grip on the Royal Society by excluding his detractors and stacking it with loyal supporters. When Whiston, by then an estranged friend, was proposed for membership in 1720 Newton threatened to resign if he were elected. Predictably his gambit succeeded. It has even been claimed that he strategically had foreign natural philosophers elected to the Royal Society on the tacit understanding that they'd do everything in their power to replicate the *experimentum crucis*. Whatever the truth of this accusation, disagreeing with Newton was unquestionably an imprudent career move that few could afford to make. So when Rizzetti launched his attack on Newton during the 1720s, he was in effect a Lilliputian facing a well-connected giant. Perhaps this is why one can detect fear behind his introductory statement: 'is it not right that I should speak of these things?'

Of course, not everyone stuck to Newton through the instinct of self-preservation. Such were his accomplishments that most were prepared to take a great deal on trust. But in the case of his theory of light Newton did stifle debate and impose a premature uniformity of opinion. When, around 1710, he became known as the 'best broacher of light', this owed much to his skill in manoeuvring critics into the shade. But this is not something we should be surprised about. Uncomfortable though the idea may be, status and prestige are as important in producing conformity in the scientific domain as they are in virtually every other walk of life. Nor is the exercise of power invariably detrimental to the progress of science: deference to the leaders of scientific institutions can prevent the inexperienced wrecking their future careers by espousing half-baked ideas.

But we must also recognize that such power can be abused. At least in the short term, the experimental method doesn't always guarantee 'Mathematical plainness'. And when cutting-edge scientists cannot agree on how to interpret an experiment's results, for a time there's scope for one side to impose its will through a combination of force of personality and institutional kudos. Unfortunately, there's no guarantee that might will always be right.

There is no question that Isaac Newton was a genius. Yet, even if he treated them as such, as Schaffer has revealed, his rivals in the debate over the nature of light and colour were certainly no fools. Given how little was then known about the qualities of prismatic glass, it made perfect sense for Lucas, Mariotte, and Rizzetti to claim that the *experimentum crucis* served only to demonstrate a curious property of impure prisms. Considering the heavy-handed tactics Newton employed to refute them, it's fortunate for his posthumous reputation that he happened to be correct.

Engraved by J. Wright,
From a Picture by S.r Geo. Chalmers Bar.t 1783.

JAMES LIND, M. D.

Physician of Haslar Hospital.

Dr James Lind and the Navy's scourge   5

James Lind (1716–1794)

Lind had battled [unsuccessfully] against confused
history, ignorance, and bureaucratic prejudice to
clarify the practical means of bringing to an end one
of the most destructive diseases . . . James Lind was
clearly the victim of a conspiracy.

David I. Harvie, *Limeys* (2002).

At the outbreak of the Austrian War of Succession in 1740, Commodore
George Anson, captain of HMS *Centurion*, was sent to the Pacific at the
head of a small fleet. His mission was to capture a Spanish treasure galleon.
But for Anson and his crew it was the beginning of a hellish odyssey. Some
medical texts of the time warned that crews should 'eat very little' of
'oranges, lemons and pineapples' since 'they are the commonest cause
of fevers.' Partly in response to this lethal advice, when Commodore
Anson's crewmen developed the open sores, swollen gums, and loosened
teeth typical of scurvy, the ship's surgeons dispensed weak solutions of
sulphuric acid. This was worse than useless.

Four terrible years later makeshift headstones all along the Pacific
coastline told a sorry tale: only 145 of the original complement of 1000
men were still alive. Nearly all the rest had succumbed to the malady
dubbed by one of Elizabeth I's sea captains 'the plague of the sea'. Towards
the end of Anson's voyage there were only enough survivors to man a
single ship. And even though this lone vessel still managed to capture the
Spanish bullion, the voyage will always be remembered more for its
mortality than for its bounty. 'The land is man's proper element,'
concluded one traumatized lieutenant on his return. But just three
years later, with hostilities drawing to a close, an event took place that is
now celebrated as the turning-point in mankind's long battle against
scurvy.

LEFT: *James Lind © The Wellcome Library, London.*

In 1747, the naval surgeon James Lind was aboard the man-o'-war HMS *Salisbury* while it helped to enforce a blockade of Continental ports. Lind seized the opportunity to separate twelve seamen whose severe joint pains, rotting gums, foetid breath, malaise, and haemorrhaging left no doubt that they were dying of scurvy. He then split his dozen men into pairs, put them all on a basic diet, and treated each pair with a different remedy held to be effective against scurvy. By testing each of these supposed 'anti-scorbutics' separately, James Lind was able to make an important deduction: that scurvy is curable with the simple addition of citrus fruits to seamen's diets. Lind had just performed the first controlled clinical trial in recorded history. More than this, he'd shown the way to eliminate one of the all-time biggest killers of sailors. 'This discovery,' wrote the biographer Louis Roddis, 'is one of the great chapters in all human history.'

That James Lind performed the first controlled clinical trial and that he deduced from it the value of citrus fruits in treating scurvy is beyond dispute. But this is only the beginning of the legend that has grown up around his name. The story told in dozens of books and articles portrays a hero of truly epic dimensions, who battled heroically against narrow-minded prejudice for much of his life, desperately trying to persuade the Admiralty to supply naval vessels with fruit and vegetables. Lind tragically died, these writers gravely reflect, without achieving the recognition he deserved, and not until forty years after his ground-breaking experiment did the Admiralty acknowledge the true value of his results.

On the face of it this seems a fair reconstruction of events. After all, only during the 1790s did the Admiralty properly recognize the anti-scorbutic properties of citrus fruits. Even as late as March 1786 its Office for Sick and Hurt Seamen had noted 'lemons and oranges [are] of no service, either in the prevention, or the cure of that disease.' Moreover, it took the tireless efforts of another Scotsman, Sir Gilbert Blane, during the closing years of the century, to drive home at last the merits of stocking ships with large quantities of fruit and vegetables. The effects of doing so were revolutionary. Within just a few years, scurvy all but vanished from Britain's fleets and seamen's hospitals.

For most writers there seems no difficulty in understanding why

James Lind's HMS *Salisbury* experiment was ignored: he suffered, as do nearly all great thinkers and doers, from having to live amid ordinary men and women who recoil at originality, no matter how many lives might be saved. Lind, they enthuse, was a classic lone genius only vindicated once the rest of his profession had had time to wake up from their slumbers. Reflecting on the thousands of lives lost because Lind wasn't heeded earlier, generations of historians have damned his contemporaries for 'high-minded disregard', 'callous indifference', a lack of 'moral fibre', 'negligence', and plain, unvarnished 'stupidity'. The latest biographer of Lind has gone further still. He declares that neither indifference nor bland conformism is an adequate explanation for the intransigence of the British Admiralty. James Lind, he writes, 'was clearly the victim of a conspiracy'.

But as the previous chapter on Isaac Newton shows, hindsight can be a treacherous ally to the historian. If we go back and explore the debates about scurvy during the 1700s, we see that there's a large gulf between the historical Lind and Lind as portrayed by later biographers. Most accounts assume his results to have been so unambiguous and transparently important that it took a corrupt, or at least obtuse, Admiralty to foil his attempts to make long voyages safe for seamen. On the basis of research conducted by the historians Kenneth Carpenter, William McBride, Michael Bartholomew, and Christopher Lawrence, this chapter argues that such an interpretation is almost wholly misconceived.

So seductive to the unwary scholar is the romantic platitude of the embattled hero that most modern renderings of Lind's career have crudely distorted what he actually said. James Lind, it will be seen, was undoubtedly an admirable man. But immersed as he was in an eighteenth-century medical paradigm, not even he could see the true significance of his 1747 trial. Indeed, far from being pushed out into the cold, Lind did more than anyone to undermine the value of his now famous shipboard experiment.

What happened in 1747?

Born in Edinburgh in 1716 to a well-heeled merchant family, Lind entered the Royal Navy as a surgeon's mate at the age of 23. For an ambitious young man it was fortuitous timing. War had just broken out against Spain and Lind saw plenty of action in the West Indies, the Mediterranean, in the waters off the African coast, and in the English Channel. Before long, he was fully battle-hardened and experienced in performing amputations under fire. With the naval ranks thinned by grapeshot and disease, he was also quickly ascending the ladder of seniority. Promoted to surgeon, he joined HMS *Salisbury* in 1746 and after three months at sea found himself dealing with a major outbreak of scurvy.

The *Salisbury* had long since run out of fresh provisions. Its crew was subsisting on the standard maritime fare of salted meat, hard ship's biscuits, and regular rum rations. Scurvy, not surprisingly, struck hard. Within days 80 of the 350 strong crew had been confined to their hammocks below deck, tormented by pains in their joints and their skin breaking out in ulcers. Luckily, however, before there had been more than a handful of deaths, the *Salisbury* returned to port, where its crewmen's diets temporarily improved and scurvy ceased to be a problem.

The following summer she set sail again and, after a few weeks, her sailors once more succumbed to the ravages of scurvy. It was now that Lind carried out his famous experiment. The dozen sailors he selected were 'as similar as I could have them' and he confined them to the same quarters. They were then divided into pairs and for two weeks each pair was put on one of the following regimes: the first drank cider, the second elixir vitriol (a solution of sulphuric acid), the third vinegar, the fourth seawater, the fifth a concoction of garlic, mustard seed, and 'balsam of Peru', and the sixth, two oranges and one lemon for six days, after which 'the supply was exhausted.' Two weeks later, the last pair had made almost miraculous recoveries: one was fit for duty and another well enough to attend to his sickening mates. Although, unlike the remaining pairs, those who had drunk cider were somewhat improved, this first ever controlled medical trial seemed to have proved beyond all shadow of doubt that, as Lind put it, 'oranges and lemons [are] the most effective remedies for this distemper at sea.' Soon after the completion of this experiment, the war came to a close, Lind disembarked for the last time, and he returned to Edinburgh to qualify as a physician. Half a decade later, in 1753, his *A Treatise on Scurvy* finally appeared in print.

In subsequent years, as physician to the Navy's Haslar hospital, Lind worked to advance his scurvy research. He spent years trying to identify the best anti-scorbutics, testing sailors with 'the juice of scurvy-grass, the Peruvian bark in large quantities, infusions of berries, stomach bitters . . . plums, apples and currants'. He also laboured to devise a means of rendering his lemon and orange juices down into a concentrated form that would last for several months, even years, at sea. Given all these efforts, and his powerful position in the Admiralty medical wing, it does seem surprising that Lind won so little credit. To modern biographers, as we've seen, this only serves to underline his esteemed status as a hero of science. Yet, however appealing a story this makes, it's not a view supported by the evidence. In terms of getting noticed, James Lind's main problem was that he wasn't saying anything distinctively new.

A man of no great distinction

To his credit, Lind never claimed (as some historians have done on his behalf) that he was the first to discover the value of fresh fruit in treating scurvy. Part III of his 1753 treatise ranged over the previous two centuries of naval medicine and cited scores of examples of ships' crews being saved from utter disaster by the timely provision of fresh fruit. Nor, as again Lind conceded, was he the first to recommend that ships carry fresh fruit as a preventive. As early as 1601 Sir James Lancaster, while commanding the first fleet sent by the East India Company to trade with Sumatra, had protected his men against scurvy by insisting on them receiving 'three spoonfuls' of 'the juice of lemons' every morning. But it was the Dutch East India Company that made the most of fresh fruit's properties. During the early 1600s they planted gardens and orchards at Mauritius, St Helena, and the Cape of Good Hope, where their ships regularly restocked with fruit and vegetables. By 1661 the Cape orchard comprised over 1000 citrus fruit trees.

Fruit and vegetables were still widely regarded as useful anti-scorbutics during the 1700s. Indeed, Lind's remark in his treatise that 'their experienced virtues have stood the test of near 200 years' suggests that he was quite convinced of their efficacy even before undertaking his trial. In Lind's day, fruit and vegetables were used far more sparingly than they had been in the previous century. But even if many physicians lauded fresh air, vinegar, sulphuric acid, and physical exercise above fruit and vegetables, plenty still stood by the value of these last two as anti-scorbutics. When Commodore Anson's crewmen began dying in their scores, his ships dropped anchor near the island of Juan Fernandez, and the Commodore himself helped carry the sick sailors to the shore. Regardless of what had originally dissuaded his surgeons from taking fresh provisions on board, they now plundered the island for 'the vegetables esteemed for the cure of scorbutick disorders'.

Evidently, the lessons of seventeenth-century seamen had not been entirely forgotten. Indeed, the treatment of scurvy with fruit and vegetables was still being recommended in standard medical texts. The English physician, John Huxham, for instance, celebrated in his 1757 *An Essay on Fevers* the 'surprising Things in the Cure of very deplorable scorbutic Cases' achieved by giving seamen 'Apples, Oranges, and Lemons, alone'. And since, Huxham added, 'what will cure will prevent,' he also advised naval surgeons to dispense oranges and lemons as preventives in the manner adopted a century and a half earlier by the Dutch and English East India Companies.

In 1768 Nathaniel Hulme's *Proposal for Preventing the Scurvy in the British Navy* appeared. He argued forcefully that lemon and orange juice ought to be given to seamen on a regular basis at one-third the level required to cure the disease. He also proposed means of preserving the freshness of the juices and furnished details of the prices, sources, and varying qualities of citrus fruits, all vital information entirely absent from Lind's own treatise.

Several of the expeditions fitted out by the Admiralty during the second half of the century also indicate that Lind's ideas were neither heterodox nor ignored. In June 1764 John Byron was sent to the Pacific with explicit instructions to purchase all the fresh vegetables he could at every port visited. Scurvy broke out on several occasions, but each time the acquisition of vegetable food brought it under control. On Byron's successful return, a Captain Wallis was despatched to the Pacific. He was ordered to try out three elaborate anti-scorbutics: a malt preparation called wort, a 'portable soup' comprising meat and vegetables boiled down into glue-like cakes, and a 'saloup' made from dried orchid roots. Wallis was of the opinion that both soup and saloup were made according to Lind's own instructions. This was probably a misconception, but he nevertheless returned and reported to the Admiralty that the addition of extra green vegetables and berries to the soup had cured numerous cases of scurvy. Healthy portions of sauerkraut had also, Wallis claimed, saved his men from disease. Most important of all, he added that he'd made a detour to the East Indies in order to pick up 500 limes for each man and several hundred more that were left around the deck to be added to water at will. He returned having lost not a single man to scurvy.

During his equally mortality-free 1768 expedition to the South Seas, the exceptionally gifted Captain James Cook was just as assured of the benefits of fresh fruit and vegetables. The holds of HMS *Endeavour* were packed with nutritious foodstuffs, and at every conceivable opportunity Cook or his quartermaster went ashore to purchase vegetables that his men were commanded to eat. The surgeon's mate aboard the *Endeavour* enumerated the full range of anti-scorbutics 'my Lords Commissioners of the Admiralty were pleased to order to be put on board': 'Sour Kraut, Mustard, Vinegar, Wheat, Inspissated Oranges and Lemon Juices, Saloup, Portable Soup, Sugar, [and] Molasses'. As the list reveals, no one concerned seems to have been prepared to stake their reputation, and the lives of dozens of sailors, upon a single remedy. But it's also clear that fresh fruit and leafy vegetables were among the leading contenders.

On Cook's next voyage, departing for the southern Atlantic in July 1772, his ships' stores again contained large quantities of sauerkraut and salted cabbage, portable soup, malt, saloup, mustard, and this time soda water too. Cook also made a point of ordering another few casks of the extract 'of oranges and lemons which we found of great use in preventing the scurvy from laying hold of our men'. Cook's second expedition returned to Britain without a single death due to scurvy. He almost certainly never read Lind's treatise, and it's simply untrue that he gave his men daily rations of lemon juice, but his insistence on taking 'inspissated oranges and lemon juices' and collecting fresh vegetables from local ports underscores just how mainstream Lind's ideas were. Far from being shipwrecked by the publication of his treatise, Lind had jumped aboard a vessel that was already moving, albeit slowly, in exactly the right direction.

Nevertheless, Lind had performed an exceptionally elegant experiment in 1747 that might have been expected to draw *special* attention to both himself and to the value of lemon and orange juice. At first glance, the quality of his results does not seem consistent with the fact that Captain Cook's expeditions were only equipped with a meagre few barrels of orange juice, mixed with brandy as a preservative. For Lind's modern admirers, this failure to take proper note of his controlled trial can only be explained in one of two ways: indifference or corruption. So how valid are these weighty charges?

Friends in high places

We can quickly dispose of the first accusation. The expeditions of Byron, Wallis, and Cook demonstrate that the Admiralty went to considerable effort and expense in trying to find out which of the various anti-scorbutics actually worked. This gives the lie to the notion that Lind was ignored because the authorities simply didn't care about the loss of seamen, who could be replaced by the simple expedient of sending out more press-gangs. For a nation that depended on efficient trade, the safe delivery of slaves to the New World, and the rapid deployment of troops to distant colonies, scurvy presented at least as big a threat as enemy fleets. Lind was quite right in pointing out that the British Navy 'lost far more men to disease than to the sword'. Finding effective anti-scorbutics was among the highest of the Admiralty's priorities. And the Lords of the Admiralty felt that they were maintaining an open and critical mind in doing so.

It's also time to bury another, equally cynical, explanation for Lind's failure to achieve proper recognition: that his reputation was destroyed to serve the political interests of his better-connected opponents. The first thing to notice here is that Lind was in anything but a marginal position. In 1753 he'd played the master-stroke of dedicating his treatise to

Commodore (by then Lord) Anson. Still haunted by the deaths of over 80 per cent of his crew between 1740 and 1744, Anson seized upon Lind's ideas with alacrity. And, as First Lord of the Admiralty, Lord Anson made sure that Lind was rewarded with the prestigious appointment of Physician to the Haslar Naval Hospital near Portsmouth. This was one of the largest and newest of England's naval hospitals, and there Lind presided over 2000 patients. An appointment to such an important medical benefice hardly amounted to ostracism. On the contrary, when and if he cared to raise it, Lind had acquired an influential voice in Admiralty affairs.

Even so, several writers have argued that two physicians, namely Sir John Pringle and David MacBride, deliberately suppressed Lind's ideas. Since these two men unquestionably did stand in the way of the acceptance of Lind's ideas, their rival claims need to be looked at. In 1750 Sir John Pringle announced that the state of scorbutic tissues and organs proved that the disease involved the body undergoing putrescence, a claim well supported by the open sores, gum erosion, and loss of teeth displayed by any patient suffering from scurvy. From this, Pringle deduced that anything that prevented rotting would also provide a fail-safe remedy for scurvy. The vital question was how the process of putrefaction could be reversed. Pringle undertook numerous experiments to find out. Several of these studies seemed to suggest that in the presence of a fermenting substance, putrefaction no longer takes place. And, since fermenting materials release carbon dioxide, known then as 'fixed air', Pringle formed the view that anything that restored this gas to the body would inevitably cure scurvy.

In 1754 David MacBride developed Pringle's ideas. He eventually decided that malt wort was the perfect remedy. In the patient's digestive tract, he claimed, malt wort ferments and releases fixed air that is 'thrown into the blood' and thereby curtails putrefaction. In addition, MacBride boasted, wort takes up very little space and loses none of its potency during long voyages. Thus both Captains Wallis and Cook had considerable quantities of MacBride's wort in their holds. And although we now know that it only contains traces of Vitamin C, Cook was impressed by its efficacy. 'This is without doubt one of the best anti-scorbutic sea-medicines yet found out,' Cook informed the Admiralty, though he added that he was 'not altogether of the opinion, that it will cure it in an advanced state at sea'. Several years later, the famous chemist, Joseph Priestley, applauded MacBride's reasoning and invented a means of producing soda water in the expectation that it too would restore fixed air to the putrefying body. To the chagrin of his ship's surgeon, a hefty machine for preparing fizzy water was taken by Cook on his second voyage.

For many writers Pringle and MacBride are the chief villains of this story. Both are disparaged for sticking to 'primitive' medical thinking. But the main allegation against them is that they exploited their social connections to have their own bogus theories treated with undue indulgence. Lind 'had no influence with the Lords of the Admiralty', one author writes, and so could make no impression on its 'hierarchy of patronage and arrogance which was more inclined to favour society physicians and their often fraudulent remedies'. In the case of Pringle there's a small amount of truth in this claim. As President of the Royal Society, he was a powerful man whose ideas, mistaken or not, were always likely to be taken seriously. Yet there's absolutely no evidence to suggest that he ever deliberately pulled rank to have the fixed-air cure judged more effective than it actually was. His contribution to the scurvy story may have proved to be a blind alley, but there are no valid grounds for impugning his character or motivation.

The charge against MacBride is no more adhesive. Vilified by one historian for displaying 'extraordinary arrogance and persistence' and by another for 'merciless self-promotion', MacBride did indeed play the patronage game to secure a sea trial for his wort mixture. But in practice this went no further than persuading his brother to test out the supposed remedy on a trip to the Falkland Islands. Several modern admirers of Lind have claimed that, through his brother, MacBride also obtained the favour of two powerful patrons, the fourth Earl of Sandwich and Admiral Sir Hugh Palliser, who made sure that wort continued to be supplied to ships years after its uselessness ought to have been recognized. Recent research, however, shows that this is simply untrue. Not until the late 1770s, years after the Admiralty first stocked Cook's ship with plentiful quantities of wort, did Captain MacBride begin to forge links with these influential men. Even then he could not have retained Palliser's favour for very long.

In 1778, he gave evidence for the prosecution during court-martial proceedings in which Palliser was accused of 'disobeying orders' and showing alleged 'timidity in battle'. This was hardly the sort of toadying behaviour of which the MacBrides have been accused and for which they might have expected special favours.

Writers have had understandable difficulty in making sense of the popularity of MacBride's wort remedy without imputing chicanery or negligence. This is because it's hard to imagine that such a spurious scientific idea might once have seemed perfectly rational. In the late 1700s, however, the putrescence theory was at the very cutting edge of medical science. Wort continued to be supplied to ships even after several disappointing trials, not because MacBride was well connected, but because his theory appealed strongly to the mind-set of the medical establishment of the time. This was an age of prodigious advances in the science of chemistry, a field in which Joseph Priestley and Joseph Black had acquired immense reputations. And it seemed to many that the newly isolated atmospheric gases would provide the cure for most of the ills which flesh is heir to. Unfortunately for Lind and for thousands of sailors, natural philosophers had become rather carried away. Driven on by overoptimism, they'd been led to make overblown claims for the new science.

Trials and error

So Lind was neither a lone voice nor the victim of a MacBride conspiracy. Nevertheless, basic questions are still left begging. How could the Admiralty have failed to recognize the significance of his experiment aboard HMS *Salisbury*? Why did they continue trying out dozens of different anti-scorbutics when fresh fruit had worked so superbly? To answer these questions we need to step inside the mind of the typical eighteenth-century physician. This allows us to appreciate that there were really rather good reasons why they failed to see a lack of fruit and vegetables as having anything to do with scurvy.

In the first place, the routine experiences of seamen simply didn't suggest a single cause or cure for this malady. When a ship had been at sea for several weeks it ran out of fresh fruit and vegetables. But the craft was also dirtier, damper, and stuffier, and often less disciplined, than when it had left port. How could physicians tell which of these conditions actually produced scurvy? In addition, although when the ship reached a foreign port its crew generally improved in health (except for cases of venereal disease), this could have been a result of fresh air and water, drier living quarters, and the opportunity for the sailors to stretch their limbs, not just their consumption of fresher foods. No wonder that in most outbreaks of scurvy more than one cause was implicated.

This reliance on multi-causal explanations for scurvy also chimed with standard medical thinking. Physicians of the 1700s believed that nearly all diseases could be caused in a variety of different ways: it was thought that the same illness could appear in different people but for diverse reasons. So, being caught in a heavy downpour of rain would be expected to cause, say, tuberculosis in some and influenza, ophthalmia, or measles in others. It all depended on the individual's personal susceptibilities. In this millennia old paradigm, still dominant a century before the emergence of modern germ theory, there were neither necessary causes nor effects for the majority of illnesses.

The same principle applied to cures. How any given patient responded to treatment, be it bleeding, purging, emetics, or some kind of pill, potion, or tincture, was reckoned to depend upon their personal constitution, and this was thought to vary widely from individual to individual. As such, physicians felt they had to tailor their remedies to the idiosyncrasies of each patient or, as was usually the case aboard ship, all manner of treatments were prescribed in the hope that at least one would work. An unfortunate consequence of this way of thinking was that physicians weren't prepared for the possibility that substances like lemon or orange juice would *always* do the trick. Physicians of the 1700s used the word 'specific' for remedies that effected a cure irrespective of varying constitutions, but genuine specifics were acknowledged to be great rarities. In fact, the very word excited distrust among physicians who associated it with the fraudulent claims and overblown assertions of itinerant quacks. Thus, when the English physician, William Withering, discovered in the 1730s that the compound digitalis, derived from the common foxglove, is invariably effective in the treatment of certain heart conditions, fellow physicians denounced him as a shameless profiteer, a disgrace to his profession. Medicines just weren't expected to work in that way.

Expectations play an important role in determining how scientific inquiry is conducted. And because nobody thought that citrus fruits, if they worked at all, would be effective in each and every case, they failed to appreciate the full significance of Lind's 1747 experiment. For the same reasons, they could see no cause to adopt the method of the controlled clinical trial. As we've seen, the naval expeditions of Byron, Wallis, and Cook were all supplied with such a variety of anti-scorbutics that there was no way of telling which were the most efficacious. Even when contemporaries did discuss their comparative merits, they showed only the weakest grasp of the possibility of making each in turn the independent variable in a formal test. For instance, Captain Cook and his surgeon's mate Perry spoke highly of wort, but Perry felt obliged to admit, 'It is impossible for me to say which was most conducive to our preservation from Scurvy so many preservatives [being] used.' Likewise, when MacBride persuaded his brother to take large supplies of wort on his voyage to the Falkland Islands, Captain MacBride unthinkingly contaminated the experiment by giving the sailors treated with wort, apples and oranges as well. Perhaps most strikingly, when lemon juice was finally made a regular issue in the British navy, Sir Gilbert Blane and his Admiralty colleagues remained absolutely convinced that it could only eradicate scurvy if accompanied by strict discipline and scrupulous cleanliness aboard ship.

So the medical mind-set of the late 1700s still made it very hard for physicians, not to mention the Admiralty, to accept that citrus juice could provide a full cure on its own. But what was James Lind's own view? Did he manage to escape the coils of these conventional ideas to establish citrus fruit as a true specific? Absolutely not. And this is where the story taught in textbooks and popular accounts most seriously founders. Lind was unequivocal on this matter: '[T]here is not in nature to be found,' he noted in his *Treatise*, 'an universal remedy for any one distemper ... so in the scurvy, deviations from the general method of cure become often necessary, according as particular symptoms of distress present themselves.' Incredible as it may seem, given the thrust of decades of hagiography, Lind did *not* believe fresh fruit and vegetables to be true specifics.

This scepticism about the anti-scorbutic properties of fruit and vegetables is hard to square with Lind's beautiful HMS *Salisbury* results. But Lind probably never saw these results as conclusive. After all, sailing lore was replete with accounts of vessels that had been at sea for months at a time with no fresh food and yet had managed to escape scurvy altogether. On land, too, there were lots of people said to have remained free of scurvy, but who consumed hardly any fruit and vegetables. Cases such as these, according to Lind, refuted the idea 'that health and life cannot be preserved long, without the use of green herbage, vegetables, and fruits; and that a long abstinence from these is alone the cause of the disease'. Lind also claimed that on at least one occasion aboard HMS *Salisbury*, scurvy had broken out long before the vegetable supply was exhausted. And so, while Lind always attached some value to his sea trial, he never felt entitled to conclude that it was of overwhelming significance. The clarity of the results obtained was contradicted by dozens of other credible reports and observations. Lind would have been very surprised to learn that 1747 would one day become such a signal date in the timeline of the history of medicine.

But there's another reason why Lind failed to see his HMS *Salisbury* results as in any way decisive and it too arose out of prevailing medical ways of thinking.

I have a theory . . .

Physicians in the 1700s, still inspired by the mechanical philosophy of René Descartes, often saw the living body as little more than a machine comprising the biological equivalents of pumps, pulleys, valves, pipes, bellows, and furnaces. The smooth operation of this delicate contraption was said to depend on the proper regulation of its various intakes and products. Too much of certain foods and liquids was a commonly cited cause of ill health. So was inadequate excretion. In this paradigm, some physicians believed scurvy to be the result of blockages in the body's plumbing apparatus that prevented it from properly expelling waste.

The unwanted substances might be corrupted blood, bile, phlegm, sweat, urine, or faeces. But whatever they were, their progressive build-up was said to poison the body and thereby induce ill health.

Lind's own explanation for why fresh fruit and vegetables were often effective as anti-scorbutics slotted neatly into this paradigm. In his 1753 *Treatise* he explained that the humid conditions aboard ship reduce sweating and cause the skin's pores to become choked up. This in turn led to an increase in the retention of 'noxious' substances. No longer expelled through the skin, these toxins built up and, if nothing was done to evacuate them, they damaged the body. Fruit and vegetables solved the problem, at least some of the time, Lind explained, by combining with other bodily fluids to produce a soapy mixture. The resulting lather broke down the 'excrementitious humours' into tiny particles that could be more easily expelled through the pores.

This understanding of the nature of scurvy, comfortably nested in standard medical theory, had major implications for Lind's ideas as to its causes and treatment. His view of the disease as the result of sweat trapped in the body implied that the 'principal and the main predisposing cause' of scurvy was not poor nutrition but excess humidity. Accordingly, too much moisture was identified as the chief culprit of shipboard scurvy in Lind's 1753 *Treatise*. More important, his theory of the cause of scurvy told him that while fruit and vegetables might act as useful anti-scorbutics, there were numerous other ways to combat scurvy. 'Plenty of recent vegetables' were to be consumed, Lind wrote, but 'only if they can be procured'. If they weren't available, all was not lost.

Keeping scurvy at bay required, above all else, making sure the sailors' pores stayed open and perspiring. The key here was regular exercise. There are 'numerous instances', Lind noted, of patients making 'a perfect recovery' from scurvy in its early stages, who had carried on eating exactly the same hard ship's tack but were 'able to use due exercise'. For Lind this made sense because intense physical activity stimulated sweating and so promoted the expulsion of 'excrementitious humours'. He also observed that the crews of 'small and airy' ships were much less susceptible to scurvy than those in 'large and crowded ships'. Ventilation increased the efficiency of perspiration. The chief responsibility of the ship's medical officer was therefore to get the sailors sweating.

Lind had all sorts of ideas as to how this could be achieved: through regular fumigation of the forecastle with burning tar, providing clean, dry bedding, as well as by encouraging proper exercise. A friend of technology, Lind suggested that ships acquire a Sutton Ventilating Machine and even a contraption designed to simulate energetic horse-riding. Anything was worthwhile, so long as it increased sweating. Lind did stipulate other healthful conditions, including a happy and relaxed state of mind. He also advised surgeons to encourage other forms of excretion through purging and rubbing olive oil upon the belly to promote urination. But the primary preventive for him was a 'dry, pure air' combined with a diet of 'easy digestion'. And the 'first step' a ship's surgeon or captain had to take when scurvy broke out was to provide 'a change of air'. The same went for landlubbers who succumbed. It was imperative that they 'remove into dry, chearful, and better-aired habitations'.

To James Lind, scurvy-free ships were those with properly managed environments. If the crew was made to exercise regularly and sailors exposed their skin to warm sea breezes, then one would not expect scurvy to strike. If it did, during the early stages the outbreak could be arrested through drying out the air and increasing ventilation. Only when this failed, presumably because the patient's pores were fully clogged up, did the medical team need to think seriously about the seaman's diet. Even then, Lind didn't claim that fruit and greenstuff was the only anti-scorbutic cuisine. The same 'saponaceous attenuating, and resolving virtue', he argued, is also to be found in unleavened bread, vinegar, onions, and cabbages 'pickled in the Dutch style'. 'I do not mean to say that lemon juice and wine are the only remedy for the scurvy,' he later added, 'this disease, like many others, may be cured by medicines of very different, and opposite qualities to each other.' James Lind, in short, did not concentrate his attention on substances later shown to contain the true anti-scorbutic, Vitamin C.

Sending mixed messages

All this made complete sense to anyone committed to the theory that disease was nothing but the build-up of noxious fluids. So strong was

Lind's commitment to the dominant theory of the day that he failed to recognize that citrus fruits were a genuine specific. Instead he proposed an entire buffet of anti-scorbutics. And even if this reflected neither faulty logic nor the hedging of bets, in terms of getting the authorities to accept the value of citrus fruit it was a disaster.

The problem was that the Admiralty was reluctant to supply fresh fruit and vegetables if anything else would do the same job. To those provisioning His Majesty's ships, fruit and vegetables had two serious drawbacks: their perishability and what Sir John Pringle referred to in 1776 as their considerable 'dearness'. In order to convince the Admiralty that oranges, lemons, and fresh vegetables were worth every shilling and more, Lind needed to extol their powers with unrestrained eloquence and dismiss the alternatives as utterly worthless. Unfortunately, because Lind did not believe fruit and vegetables to be the only anti-scorbutics, this was not a strategy he was ever likely to adopt. Instead he blithely diluted what to us is the key message contained in his 1747 experiment.

In consequence, anyone in the Admiralty who managed to wade through the 400 pages of Lind's meandering and stultifying prose would have been able to draw from it at least modest support for the cheaper and more manageable practices that Byron, Cook, and Wallis employed. Several senior members of the Admiralty were struck by the results of the HMS *Salisbury* experiment. But as even Lind believed that other methods worked, it was felt to be enough for surgeons to place reliance on improved ventilation, plus a few gallons of fruit juice as part of their medical supplies for use when all else failed. The inescapable conclusion of all this is that if anybody took the lead in undermining the basic message of the HMS *Salisbury* experiment, it was Lind himself.

It would take an entirely unplanned, naturalistic experiment to make the Admiralty see the true worth of citrus fruits. In 1793 scurvy began to cut through the Mediterranean fleet assembled under the command of Lord Hood. Battle was soon to be joined with the French fleet and there was no possibility of returning to shore. Then the chief physician had a bright idea. He sent a vessel into port 'for the express purpose of obtaining lemons for the use of the fleet'. It was a sterling success. Within days the sailors' sores had healed, their ulcers had begun to die away, and an 'order was soon obtained that no ship under his lordship's command should leave port without being previously furnished with an ample supply of lemons'. Accidental experiments such as this eventually convinced the naval physician, Sir Gilbert Blane, that 'Lemons and oranges . . . are the real specifics.'

With Blane's backing, the Navy's Hurt and Sick Board finally decided to give citrus juice, used alone, a decent trial. On the voyage of Admiral Gardner's HMS *Suffolk* to Madras, each seaman was given a glass of lemon juice, sugar, and grog every day. That there was not a single case of scurvy on board during a passage lasting twenty-three weeks convinced most that the trial had been a success. Ultimately, sheer weight of experience forced the Admiralty to see that citrus fruits were substantially more effective than any of the other anti-scorbutics on offer. Even then, most physicians continued to see them only as the most important weapon in a larger battery of anti-scorbutic measures.

All quiet on the Haslar front

All this time, until his death in 1794, Lind pursued his scurvy research at Haslar naval hospital. According to some historians, these were intensely frustrating years which he spent battling tirelessly but thanklessly against the dogmatic conservatism of the Admiralty. In reality, tenacious striving was just not Lind's style. He made no attempt to draw up a set of specific dietary recommendations that the Navy's Sick and Hurt Board could have implemented. His fellow naval physician, Nathaniel Hulme, explicitly advised the Admiralty to supply ships with enough citrus juice for each man to be able to drink an ounce and a half every day. Hulme also insisted that its provision be the responsibility of the government. Lind, in contrast, directed his suggestions to the sailors themselves, cautioning them to carry onions and have concoctions of citrus fruits prepared for their personal use at every port.

Diffidence, though, wasn't the only reason Lind became a somewhat marginal figure. His chief problem was that for the remainder of his life he struggled to reproduce the clear and unequivocal results he'd obtained aboard HMS *Salisbury*. At Haslar, Lind repeatedly ran up against the fact that nature is far too complex for just a few experiments to reveal her secrets. On dozens of occasions he confined scorbutic sailors to private rooms and regulated their diets. Yet now his results never seemed to be clear-cut. Patients deprived of vegetable matter stubbornly refused to develop symptoms of scurvy, whereas those with the disease already improved on what he took to be the least nutritious of diets. Definitively explaining these results is no longer possible, but it seems likely that because Lind didn't know that the anti-scorbutic principle is found in a wide variety of foodstuffs, he was inadvertently giving all his patients adequate doses of Vitamin C.

Lind's biggest disappointment arose from his attempt to produce a concoction of lemon and orange juice that wouldn't spoil on long journeys or take up too much space in the ship's hold. The result was called a 'rob' and it was made by squeezing oranges and lemons into large bowls that were held just beneath boiling point for twelve to fourteen hours, or until the juice had evaporated down to a syrup. 'Thus the acid, and virtues of twelve dozen of lemons or oranges,' Lind wrote, 'may be put into a quart-bottle, and preserved for several years.' This at least was the claim. But, as is so often the case in science, a seemingly rational idea ran aground because of what the scientist could not know. Heating his fruit juice for such a long period of time so damaged the Vitamin C it contained that its potency was reduced by almost 90 per cent. As a result, it became harder than ever to prove the efficacy of citrus fruits. No longer would just a few drops over the course of several days result in seemingly miraculous recoveries. And far from the rob retaining its potency for years, modern studies suggest that by the end of the first month it would have lost three-quarters of its already diminished nutritional value.

The poor performance of the citrus syrup seems to have badly shaken Lind's confidence. In the Introduction to the revised edition of his treatise, published in 1772, he complained, 'though a few partial facts and observations may, for a little, flatter with hopes of greater success, yet more enlarged experience must ever evince the fallacy of all positive assertions in the healing art.' It wasn't a confident opening; nor did it invite further reading.

By the time of the third and final edition of his *Treatise*, published in 1777, Lind was even less sure that anything definitive could be said about the cause and the cure of scurvy. He now concluded that scurvy could strike even those who 'live upon fresh greens and vegetables, or the most wholesome diet, and in the purest air'. And he expanded on an already bewildering array of recommended anti-scorbutics, attaching particular importance to cream of tartar, which we now know to be useless. In fact, Lind suggested so many treatments that, as the historian Christopher Lawrence has noted, leading medical writers of the 1780s and 1790s cited material taken from his *Treatise* to support the claim that fresh greenstuff does not cure scurvy at all.

Getting Lind right

By projecting back to the late eighteenth century what seems obvious to us today, writers have misrepresented the life of James Lind, but his is a story that has much to teach us about the difficulties of scientific discovery. Having performed a brilliant experiment, neither he nor his contemporaries were able properly to interpret it because of the medical paradigm in which they were immersed. He wrote a book avowing the value of citrus fruits, but then weakened his message by also recommending regular fumigation, physical exercise, fresh air, and a variety of other foodstuffs. He manufactured a syrupy extract of oranges and lemons that, far from concentrating the anti-scorbutic element, instead severely damaged it. In addition, for many years his ideas struggled to compete with a chemical theory of the causes and cure of scurvy that many of his more scientifically inclined peers found simply irresistible. Lastly, rather than being the first to recognize that scurvy is a nutritional deficiency disease, as writers still routinely claim, Lind explicitly attacked this view in arguing that citrus juices do no more than help the body expel noxious fluids that otherwise build up and cause putrescence.

The Lind we are left with is still an impressive figure, but the emphasis historians have tended to place on his 1747 sea trial can now be seen as grossly exaggerated. The HMS *Salisbury* experiment was no more important to Lind himself than dozens of other observations and, for what seemed to him very good reasons, he made no attempt to present it as a definitive comparison of the rival anti-scorbutics. In light of this it's clear that the Admiralty does not deserve the aggressive censure it has received from many of those retelling the story. Considering the ambiguity of Lind's findings and the ineffectiveness of his 'rob', its four-decade delay in imposing the consumption of citrus fruits aboard ships of the line can hardly be wondered at.

But for students of history, the most important lesson to be drawn from this re-analysis is just how careful we need to be in reflecting on past scientific views. To do justice to our ancestors we have to put modern knowledge to one side. Only then can we begin to see how paradigms now long since rejected once made sense of a complex and ever-changing world. Of course, the firmness with which scientists held theories later shown to be invalid infuriates the modern reader, especially when lives were at stake. But in the absence of key pieces of evidence, false ideas can have a beguiling plausibility. This has an obvious bearing on the present. It is a fundamental tenet of modern science that all theories are at best provisional. Accordingly, we have to accept that however satisfactory our current scientific theories may appear, in a century's time many will have been proven wrong and others will seem quaintly incomplete. If this statement reflects the relentless progress of science, it also reminds us of the inescapable limitations of human knowledge.

Ignaz Philipp Semmelweis (1818–65).

Ignaz Semmelweis (1818–1865)

It is the doctrine of Semmelweis which lies at the
foundation of all our practical work today . . . The
great revolution of modern times in Obstetrics as
well as in Surgery is the result of the one that,
complete and clear, first arose in the mind of
Semmelweis, and was embodied in the practice of
which he was the pioneer.

Sir William Sinclair,
Semmelweis: His Life and Doctrine (1909).

The traditional account of the life of Ignaz Semmelweis provides us with
the supreme example of the lone genius destroyed by jealous and ignorant
opposition. The Semmelweis story has become a classic of the genre for
two main reasons. First, the process of destruction culminated in the
death of our hero. Second, rather than some abstruse aspect of natural
science, Semmelweis devoted his life to the very practical end of reducing
the horrendous death rate among nursing mothers. It's a stirring saga in
which the extreme vulnerability of those Semmelweis tried to save
interplays powerfully with the pathos of his own death and apparent
failure. Yet it all began so well. In May 1847 Semmelweis had a bowl of
disinfectant placed in the Vienna Hospital's maternity ward in which
he worked. He then ordered the staff to wash their hands in it before
touching expectant mothers. The effect of this simple reform was
extraordinary. Over the following months, the number of women dying
from childbed fever plummeted and hundreds of children were saved the
misery of a motherless childhood.

As the standard model of scientific heroism requires, Semmelweis'
tremendous achievements then went almost entirely unrecognized. Other
obstetricians ignored his results. He was victimized by his boss at the

Vienna Hospital, and, after a few years, he was forced to leave Vienna. The hospital then largely rescinded his practice of hand-washing and endemic childbed fever once more held sway. Returning to his native Hungary, Semmelweis encountered similar opposition. Once again, leading obstetricians declared hand-washing a futile waste of time and resources. Beset with criticism, he gradually slipped into insanity. In July 1865 Semmelweis' family were obliged to confine him to an asylum, where he died two weeks later from, it appears, blood poisoning. Semmelweis' plunge into oblivion was now so fast and so complete that it almost appears as if a collective decision was made to expunge him from the historical record. His was a life that had gone terribly, almost unbelievably, wrong.

Then, during the 1890s, again in strict accordance with the standard model, the man who had died in an asylum, forgotten by his entire profession, was canonized as one of the most shining figures in the history of medical science. Only Louis Pasteur, Robert Koch, and Joseph Lister were able to boast larger followings than Ignaz Semmelweis. And, ever since, he has been little short of a patron saint for all those who, believing themselves to have had a brilliant new idea, suffer rejection and humiliation. Thus a contributor to the *Proceedings of the Royal Society of Medicine* wrote in 1966: 'Many men have been endowed with clear intellects and hearts full of love for their fellow men, but in the whole history of medicine there is only one Semmelweis in the magnitude of his services to Mankind, and in the depths of his sufferings from contemporary jealous stupidity and ingratitude.' The same romanticism infuses even some of the most modern histories of medicine. 'Semmelweis lost his sanity and his life in the battle against puerperal fever,' notes one contemporary historian. 'Unwilling to compromise with corrupt and ignorant authorities . . . [he] destroyed his own career and made few converts to his doctrine.' 'Despite his success,' adds another, 'Semmelweis was criticised, ridiculed, and eventually dismissed.'

Small wonder, then, that Semmelweis became one of the greatest heroes of medical science. Everything about his troubled life smacked of the lone genius who valiantly, and with a selfless disregard for his own reputation, challenges stupid, selfish, and callous opposition in his pursuit of the truth. After all, who but the most far-sighted of men could have been so universally repudiated, only to be hailed as a hero decades after his demise? It's a story that has all the ingredients of the classic romantic drama: a tragic hero, a chorus of selfish peers who would rather persecute a brave man than admit their faults, a picaresque scene in a nineteenth-century madhouse, and a fitting climax in which the hero is vindicated, albeit posthumously, and his surviving persecutors are left to rue their short-sighted folly. In short, it's unleavened Hollywood.

But was professional jealousy solely to blame for Semmelweis' fall from grace? In this chapter I draw on recent work showing that Semmelweis' story has had to be rudely manhandled to fit into the mould of the romantic hero of science. In light of research conducted by several historians, but principally K. Codell Carter, György Gortvay, Irvine Loudon, and Imre Zoltán, we're now in a position to give a rather different account of how Semmelweis came to die, unregarded, in an asylum for the insane.

Miasmas and maternity

In 1844 Semmelweis qualified as a doctor in his native Hungary. It was a remarkable achievement for a man whose father was a grocer, but this was only the beginning. He elected to specialize in obstetrics, and two years later landed a job as midwifery assistant at Vienna's lying-in hospital. At the time this was the largest maternity hospital in Europe, with dozens of beds in each of its main wards that seemed to stretch on either side towards a distant vanishing point. Not least because of its scale, this hospital had an acute problem: as many as 10 per cent of the 7000 women who gave birth annually in its wards succumbed to deadly infections of puerperal, or childbed, fever.

Although doctors did not know or accept this for another thirty years or so, childbed fever is a form of blood poisoning (septicaemia) caused by lethal bacteria invading a uterus left tender and exposed by childbirth. Days after delivery, the mother began shivering violently, her muscles trembling and her teeth chattering. Agonizing abdominal pain followed, soon after which death ensued in roughly 30 per cent of cases. The wards of all lying-in hospitals in Europe and America were haunted by the spectre of childbed fever. But while elsewhere the fever came and went, at the Vienna hospital it stayed with terrible constancy.

Naturally, doctors had firm ideas as to what caused childbed fever. Unfortunately, these were almost entirely incorrect. Some medical theorists blamed it on the unique stresses of pregnancy and childbirth, from the tightness of maternity garments and excessive supplies of blood to the sudden drop in the mother's weight following parturition or its subsequent rise due to milk production. But several decades before doctors recognized the role of microbes in producing disease, most believed that the inhalation of poisonous airborne particles caused infectious diseases like childbed fever. These noxious 'miasmas' were said to rise up from piles of decomposing refuse, streams of sewage, and carcasses left to rot outside tanneries and abattoirs. Since hospitals of the period were generally pretty squalid places, in many cases this theory seemed to fit the facts. But ignorance of germs meant that few medics believed that infections could be passed from patient to patient upon their unwashed hands, surgical instruments, the leeches used to draw blood, or the butchers' aprons stiff with bodily fluids often worn by surgical staff. As a result, medical personnel took only the most rudimentary hygiene precautions. Disinfectants and antiseptics were seldom used in European and American hospitals of the 1840s. If an epidemic of childbed fever struck a ward, doctors and nurses did little more than increase ventilation to allow any poisonous fumes to escape, and then clear away the more unsightly surgical debris.

But if miasma theory generally held sway, in the case of childbed fever there were several alternatives. And so, deeply concerned about the appalling rates of maternal mortality, during the 1850s the Vienna hospital authorities conducted a series of studies to try and establish its true causes. None of the theories tested provided a complete answer. And, in spite of plenty of anomalies, the most popular explanation for childbed fever remained the poisonous miasma.

Rise and decline

It was at this critical juncture that Ignaz Semmelweis arrived in Vienna from his native Hungary. His timing was highly fortuitous. Fresh from medical school, eager to impress, and perhaps more affected by the private tragedies daily being enacted on his wards than more seasoned midwives, he set about trying to discover the cause of childbed fever. A seemingly unremarkable change in the running of the hospital had created the ideal conditions for Semmelweis to find the answer. In 1833, what had been a single maternity unit was split into two separate wards, with patients being admitted to each on alternate days. Six years later it was decided on administrative grounds that medical students and visiting doctors should be confined to Ward One, leaving the hospital's midwives to run Ward Two unhindered. This simple alteration was to have far-reaching effects.

Within months, deaths from childbed fever were at least four times more common in Ward One than in Ward Two. This discrepancy in mortality rates was well known to staff by the time Semmelweis arrived at the hospital. In fact, women throughout Vienna knew of the disparity. As Semmelweis worked mostly in Ward One, he was often confronted with 'moving scenes in which patients, kneeling and wringing their hands, begged to be released in order to seek admission to the second section'. Ward One's midwives, as Semmelweis later reflected, were unsurprisingly subject to stinging criticism and 'disrespect'.

So it was that the Hungarian began searching in earnest for anything that made Ward One different. His tenacity in finding the truth is nothing if not impressive. Semmelweis first performed dozens of autopsies on the fever's victims from each ward. He made sure that both wards received food from identical sources and that the same people washed their linen. He even asked the priest to vary the route by which he entered the wards so that the patients in one did not see him more than those in the other. Then, after weeks of getting nowhere, he set out on a short trip to Venice with some friends. On his return he heard of the excruciating death of one of his superiors, Professor Kolletschka, whose finger had been cut by a student's scalpel during the dissection of a woman who had died following childbirth. Semmelweis' first major breakthrough was to recognize that Kolletschka's symptoms were almost identical to those experienced by the women lying on the mortuary slabs. Now he began wondering if the source of infection might be material contained in the bodies of fever victims.

After drawing dozens of blanks, Semmelweis was now in a position to make his famous observation: only in Ward One were medical students permitted to assist in childbirth, and unlike the midwives of Ward Two, these students also performed autopsies on the cadavers of women who had recently died from childbed fever. It suddenly seemed so obvious. The students' days started in the morgue, where they handled the bodies of dead mothers, both to investigate the reasons for their deaths and to practice childbirth procedures. Then they walked upstairs into Ward One to assist in deliveries and vaginal examinations. As they neither washed their hands nor changed their clothes in between, Semmelweis concluded that they must routinely convey deadly, 'cadaverous' material from the dead to the living. 'Cadaverous particles,' he later wrote, 'adhering to hands cause the same disease among maternity patients that cadaverous particles adhering to the knife caused in Kolletschka.'

No mere armchair thinker, Semmelweis now turned his attention to prevention. He couldn't prohibit autopsies, so instead he insisted on students washing their hands in disinfectant chloride of lime between autopsies and deliveries. The results were spectacular. He achieved a 70 per cent fall in the maternal mortality rate within just a few months, a drop unheard of in any other hospital. When the death rate then began to climb again with the arrival of several more students, he immediately assigned a midwife and a student to each woman in labour and displayed their names above her bed for all to see. If the mother died then no one could doubt who had failed to wash his hands with the requisite care. Due to this new policy of public shaming, the incidence of fever fell off once more. Semmelweis now had every reason to expect a prompt rise into the forefront of medical celebrity. But as we've seen, this is not what transpired.

Instead, Semmelweis' boss, the autocratic Johannes Klein, refused to listen. Semmelweis' position in Vienna had only been temporary. When it became a permanent one he applied to stay on. Klein instantly passed him over for a much less qualified rival. At the same time Klein ruthlessly blocked a call for Semmelweis' achievements to be examined by a hospital commission. Disappointed but determined to keep trying, Semmelweis remained in Vienna for a further year. Then he gave up. Returning to Budapest, he struggled to find gainful employment and ended up working as the unpaid director of a small maternity clinic. At St Rochus Hospital, Semmelweis quickly instituted hand-washing in disinfectant and, once more, rates of childbed fever rapidly declined. Then, in 1854, his professional luck improved and he became Professor of Obstetrics at the University of Pest. There, against strident opposition, he imposed the same policy of hand-washing and also insisted that bed-linen and obstetric equipment be disinfected before use. The results were no less gratifying than they'd been at the Vienna lying-in hospital and at St Rochus. Still, however, a prominent Viennese medical journal felt able to declare: 'This chlorine-washing theory has long outlived its usefulness . . . It is time we are no longer deceived by this theory.'

In 1858 Semmelweis finally published news of his results; and his book, *The Aetiology, Concept, and Prophylaxis of Childbed Fever*, appeared in print in October 1860. It was not a success. A few respectful reviews appeared, but most were of a critical nature, and several were thoroughly dismissive. And, in the following months and years, the treatise seemed to have no discernible effect on the way most doctors, midwives, and surgeons worked.

The revolution in hospital practice that Semmelweis had keenly anticipated turned into the dampest of squibs. Barely able to contain his anger and disappointment, his mental condition began to deteriorate. He raged against the injustice of his treatment by his colleagues and hurled abuse at his critics. Eventually, his family felt it necessary to have him confined. And, on 29 July 1865, he was escorted to a Viennese asylum and left there. Two weeks later he was dead of blood poisoning, a martyr to the disease that he'd spent his life fighting. But as Semmelweis' biographers lament, even now his critics expressed no regret. At the poorly attended funeral that followed a few days after his autopsy, all memories of Semmelweis seemed to follow him into the grave. Not even his family or in-laws attended. He received no eulogies in Hungarian medical journals, nor did a single obituary appear for nine years. So the question we now have to ask is why this brilliant and humane man was ostracized by those whose admiration he most richly deserved?

A *grave* faux pas?

Semmelweis, or so generations of historians claimed, never stood a chance. In 1846 he presented his colleagues with a terrible dilemma that was destined to make enemies of even his closest medical friends. They could either endorse his theory as to the cause of childbed fever or refuse to believe that hand-washing had anything to do with the decline in mortality achieved in the Vienna hospital. The first option, though intellectually satisfying, came at a high price, for it meant each surgeon and midwife confessing to having inadvertently killed hundreds or thousands of women. Semmelweis' many modern admirers make clear that his colleagues chose the less admirable route. Their oath to Hippocrates temporarily forgotten, they repudiated Semmelweis and upheld their vanity. The price paid for shoring up their professional dignity, we are told, has to be measured in the lives of tens of thousands of women and one exceptionally courageous man.

There are, as readers may have come to expect, many problems with this account. But we cannot write it off as pure myth because there's no doubt that Semmelweis' findings did cause some of his peers acute discomfort. There's even a historical precedent. About fifty years before Semmelweis' experiences in Vienna, a Scottish physician, Andrew Gordon, had been driven out of Aberdeen for blaming the spread of childbed fever on doctors and midwives. Both Gordon and Semmelweis appalled colleagues by implicating them in what was little short of an accidental holocaust of mothers. Indeed, even Semmelweis had difficulty in accepting his findings (perhaps this is why he scrutinized the route taken by the priest before reflecting on his own from the autopsy room). Having arrived at his final conclusion, Semmelweis realized with horror that his own enthusiasm for dissection was directly responsible for the sharp upturn in mortality that had followed his arrival. 'Only God knows the number of patients who went prematurely to their graves because of me,' he wrote. Gustav Adolph Michaelis, a professor of obstetrics at Kiel in Germany, couldn't forgive himself so easily. Soon after accepting Semmelweis' doctrines, Michaelis hurled himself beneath the evening train to Hamburg. The Hungarian, it seemed, had finally vindicated Napoleon Bonaparte's cruel *bon mot*, 'You medical people will have more lives to answer for in the other world than even we generals.' Few doctors were going to follow Michaelis' lead; to many, shooting the messenger would have seemed a far more attractive option.

One could hardly, however, classify this as callous indifference. It's hard to imagine that many of Semmelweis' detractors had much conscious awareness of why they were so sceptical of his ideas. Dr Bernhard Seyfert, for instance, at the University of Prague, felt that Semmelweis' claims couldn't be ignored, so he too introduced a bowl of chloride of lime. But as a student later recalled, 'Usually, only the fingertips were dipped into an opaque fluid that had served the same purpose for many days and that was itself completely saturated with harmful matter.' It seems probable that an unconscious desire on Seyfert's part that Semmelweis be proven wrong impelled him to put the hand-washing practice to the test in the most desultory fashion. In consequence, his efforts did more harm than good.

Yet if doctors can be as self-serving as the rest of us, it's simply not true that this was the only or even the primary reason why Semmelweis failed to win a fitting accolade during his lifetime. As we'll see in the next few pages, the fact that this Hungarian was driven out of Vienna, embittered and down at heel, was in no small part due to his own political and professional miscalculations.

On the home front

Morton Thompson's richly embroidered 1951 biography of Semmelweis, *The Cry and the Covenant*, made great play of its hero's shoddy treatment at the unwashed hands of Johannes Klein. What Thompson and many other writers have neglected to say is that Klein was virtually the only senior member of staff at the Vienna medical school who expressed the slightest disapproval towards Semmelweis and his ideas. During the 1850s this school was home to many of Europe's leading doctors. Several of these rising stars, Ferdinand Ritter von Hebra, Josef Skoda, Franz Hector Arneth, and Carl Rokitansky had no hesitation in congratulating Semmelweis and swearing their allegiance to him. Far from being the beleaguered outcast of standard histories, he found himself lauded by many of the hospital's brightest, at a time when the Vienna medical school ranked alongside the very best in the world.

Hebra, described as 'perhaps the most brilliant' young doctor in the Vienna hospital, was the first to publish news of Semmelweis' findings. In two articles of 1848 he elaborated the 'cadaverous matter' thesis and claimed that the brilliant Hungarian deserved a status on a par with the great Edward Jenner. Soon after, Arneth travelled to Edinburgh, where he gave leading Scottish medical men a run-down of Semmelweis' achievements. Closer to home, in January 1849 Skoda began organizing an official committee to look into the 'currently so meaningfully reduced mortality rate'. At this stage, even Klein voted in favour of the enquiry. So, on the home front at least, things had got off to a promising start. If anything, Semmelweis' problems stemmed from the very popularity his ideas achieved among certain sections of the hospital staff.

The elected committee proposed by Skoda comprised Rokitansky, Franz Schuh, and Skoda himself. All were able and promising men. Klein, however, was incensed at the make-up of the committee and, after a bitter struggle, he ensured that it was never convened. Klein's reaction was partly a result of pique: he had wrongly assumed that as the professor of obstetrics he would automatically be on the panel and was deeply offended by his exclusion.

But this was only one aspect of his motivation. To understand why Klein vetoed an investigation that would have given Semmelweis all the backing he needed we have to look beyond the walls of the Vienna hospital and onto the city's streets and squares. Biographers of great scientists often write as if the world outside the hospital or laboratory scarcely exists. They seem to assume that powerful minds rise effortlessly above the meanness and brutality of everyday existence, finding solace only in a world of recondite speculations and a devotion to humane ends. This is generally a mistake, and it's fatal to any attempt to understand the real reasons why Ignaz Semmelweis was driven out of Vienna.

Revolutions . . .

1848 was the year of European revolution. A period of political turmoil of a scale and intensity hardly seen before, it plunged virtually the entire continent into crisis and reached its peak in a series of bloody suppressions. Its immediate cause was a dire shortage of grain following the crop failures of 1846. Food prices rose dramatically while wage levels stagnated. The knock-on effects of this dismal equation were devastating. As people were compelled to spend more of their income on necessities, demand for manufactured goods fell heavily and unemployment rose sharply. From this distance of time, these events may sound like mere abstractions; but if empty bellies breed discontent, hungry towns and cities create the preconditions for revolution. During the late 1840s impecunious middle classes and gaunt-faced labourers began to look with growing envy and disgust upon the autocratic regimes of Continental Europe. The kinds of corruption, gross indulgence, and repression that could be borne with relative equanimity on a full stomach now became insufferable.

Real trouble first broke out in Paris. As the lower and middle classes recovered from the New Year festivities of 1848, simmering resentments exploded into violent protest. The citizen militia of King Louis-Philippe defected *en masse* and the Second Republic was declared towards the end of February. Now the rest of Europe followed suit. 'When Paris sneezes,' famously remarked Austria's Chancellor Clemens von Metternich, 'Europe catches a cold.' Within a month Metternich had the misfortune to be proven right. Revolutionary fervour swept through Prussia and Italy, then engulfed the Austro-Hungarian Empire. Supporters of liberal, democratic reform took to the streets of all the Empire's great cities. Vienna, Prague, and Budapest became the scenes of unquenchable public anger. The Habsburg monarchy wobbled, its loyal servant Metternich fell, and a frightened government was compelled to enact liberal reforms.

With Vienna in the hands of the victorious rebels for much of 1848, there was a brief period of democratic innovation. During these heady days, revolutionary fervour was felt everywhere, and Vienna's universities, including the medical school, played a pivotal role in maintaining the revolution's momentum. The new liberal government decreed that students could form a national guard, the Academic Legion, responsible for defending civil liberties. Meanwhile, in Semmelweis' native Hungary, the charismatic Lajos Kossuth led a mass revolt and, at the head of a large army, issued strident demands for Hungarian independence. He drew strong support from Austrian liberals.

But what does this have to do with Ignaz Semmelweis? Except in the form of a literary garnish, one will search in vain for any mention of the 1848 revolutions in most biographies. Yet, Semmelweis embraced the revolution with genuine enthusiasm. Described by a witness as 'one of the most passionate and active revolutionary fighters', he hurried to join the Academic Legion. Even months after its dissolution, he still insisted on wearing its uniform of grey trousers and a tight blue jacket with black buttons, together with a wide-brimmed hat pierced with a black feather. Foreign visitors to the medical school were bemused by the sight of him wearing this martial attire when lecturing on obstetrics. According to several reports, Semmelweis actually took part in one of the Academic Legion's most celebrated actions. On 13 March 1848, it sabotaged an imperial army destined to suppress the Hungarian revolt by destroying hundreds of tonnes of munitions. In October the militia prevented another departure of the Viennese army, though it's not clear if Semmelweis was involved this time.

Yet by mid-October the partisans of progress were fast losing ground. The main body of imperial troops had stayed loyal to the Habsburgs, and a victorious Hungarian army, *en route* to Vienna, had disintegrated into dozens of bickering cliques, incapable of marching anywhere. The Habsburgs seized their opportunity. Vienna was bombarded with cannon and the musketeers of Windischgrätz stormed the city, ruthlessly cutting down the rebels. Fighting only ceased when the Habsburg flag once more flew atop the church of St Stephen. In the aftermath, the universities' privileges were swiftly revoked and, in concert with the conservative backlashes occurring throughout Europe, monarchical absolutism was restored. In early 1849, the imperial army also quelled the Hungarian insurrection. With this final victory, the status quo of 1847 had been re-established—well, almost, for not everything could return to normal. Once the dust had settled and the democrats were disarmed, Semmelweis found his position seriously compromised.

. . . and their young

It was an excruciating piece of bad luck that Johannes Klein, effectively the master of Semmelweis' professional fate, was among the medical school's fiercest reactionaries and, since Kossuth's rebellion, a pious xenophobe with a particular animosity towards Hungarians. By all accounts rather slow-witted, fearful of his status being undermined, and especially alarmed by youthful precocity, Klein identified closely with a monarchical regime that had lost the support of large sections of the educated classes. In 1848, when he encountered Semmelweis in Ward One with a black feather protruding from his hat, he presumably filed his annoyance away for a more propitious occasion. It came with the collapse of the revolution. Klein's position was greatly strengthened in the aftermath of 1848, and he could now prosecute a private war against this rebellious young Hungarian with the full sanction of the conservative heads of the medical school. This is the context in which we need to understand Klein's hostility to Skoda's hospital commission. In a period of intense political discord, Semmelweis and his allies had assumed leading roles in what proved to be the losing side.

Remarkably, perhaps, Klein was not at first arbitrarily hostile to Semmelweis. When Skoda first proposed the commission Klein gave it his unqualified support. Unfortunately, Skoda and Rokitansky then made the grave tactical error of excluding Klein from the committee that would do most of its investigations on Klein's own wards. As passionate liberals, these men couldn't tolerate working alongside a notorious reactionary. Perhaps they also relished the opportunity of provoking him into an unseemly rage: the conservatives might be back in power, but this didn't mean the liberals had to be slavishly respectful towards them. This was a serious and ultimately tragic miscalculation. And at this critical juncture Semmelweis' popularity among the younger, progressive doctors began to work against him. Klein immediately understood why he'd been omitted from the committee and he gladly accepted the implied challenge to his authority. Skoda and Rokitansky had made Semmelweis' case into a political contest and foolishly underestimated the resourcefulness of their opponent. On the political stage the old guard had decisively shown that they could bite back; it was a message that the Semmelweis faction should have heeded.

Immediately withdrawing his support for the inquiry, Klein requested an audience with his own superiors and explained to them how the liberal elements of the hospital were seeking to undermine his primacy on his own obstetric wards. In post-revolutionary Vienna, the watchword was 'hierarchy' and any challenges to the established chain of command were too redolent of the liberal revolution to go unpunished. The hospital authorities backed Klein without hesitation. Too late, Skoda and Rokitansky realized that they'd picked a fight they couldn't possibly win. Semmelweis and his theory were now tainted by political intrigue and Klein was given every incentive for exacting revenge on his junior colleague. He wasted little time. Within months, he'd replaced Semmelweis with Karl Braun, who had the double advantage of being unsullied by political heterodoxy and having a pre-existing aversion to his predecessor.

Although Klein effectively rescinded the policy of hand-washing in Ward One, there's no evidence that he'd ever felt strongly on the matter until politics supervened. His objection was to the man who came up with the idea, not the idea itself. Ignaz Semmelweis, in short, was a belated casualty of the year of revolutions. Nevertheless, for abandoning the hand-washing practice even after it had appeared to have such salutary effects, Klein amply deserves his reputation as a poor scientist and an individual utterly unworthy of rehabilitation.

Explaining epidemics

But what of the cool reception Semmelweis' theory received in the rest of Europe, where his politics were almost entirely unknown? Here too, it's clear that much of the damage was self-inflicted. In 1847 Semmelweis had achieved a decline in mortality that was little short of spectacular. The obvious thing was to get his results into print. To the chagrin of his

Viennese allies, however, he simply could not be induced to pick up a pen. Semmelweis claimed to have an 'inborn dislike for everything that can be called writing'. The reason for this reticence isn't entirely clear, but several of his peers also remarked on his dread of violent controversy, and it's more than likely that Semmelweis had a deep-seated fear of criticism. Perhaps he already detected in himself the acute sensitivity that would one day help reduce him to a state of abject insanity.

As it was, Semmelweis had the good fortune to attract allies willing to proselytize for him. Yet they too met with incredulous, unsympathetic, and often contemptuous audiences across Europe. This scepticism may seem in retrospect ignorant and mean-spirited. But it's crucial to recognize that much of it actually arose from genuine weaknesses in Semmelweis' case. In 1847 he claimed that the single, necessary cause of childbed fever was 'cadaveric poison'. However, doctors throughout Europe immediately knew that this couldn't be right. In Britain, it just so happened that morgues were never situated in lying-in hospitals. In France they sometimes were, but most medical students practised with the technical paraphernalia of midwifery on live prostitutes rather than dead, infectious mothers. The hands of both British and French students and doctors were therefore typically free from the noxious substances implicated by Semmelweis. The Vienna lying-in hospital, in consequence, seemed to be a special case and the Hungarian's ideas to have limited application.

Even in Semmelweis' own wards the cadaveric poison theory quickly foundered. After all, although there were far fewer fever deaths in Ward Two, childbed fever did strike and yet Ward Two's staff had virtually no contact at all with cadavers. Nor could Semmelweis' theory account for the hundreds of cases of childbed fever that occurred annually beyond the hospital gates. Semmelweis eventually abandoned his original theory after a woman with 'a discharging carious left knee' was brought in to give birth. A wave of deaths in beds surrounding hers made it clear that corpses weren't the only problem. Semmelweis now realized that his cadaveric poison thesis was untenable. So he stretched his theory and argued that the animate, no less than the deceased, can generate deadly substances. He imagined these to comprise 'decomposing animal organic matter', and concluded that they caused nearly all cases of childbed fever.

This was a much better theory. But even this one had a crippling weakness. Everywhere, including the Viennese lying-in hospital, childbed fever displayed epidemic features. In Ward One, for instance, fever had always been present but at certain times of the year its incidence had usually risen sharply. This was very hard to square with Semmelweis' latest theory. For if dead tissue was the only requirement for the spread of infection, then there should have been no seasonal variation in its frequency at all.

To his credit, Semmelweis was quick to spot this flaw and he did all he could to deal with it. His attempts, however, were less than cogent. In letters to other doctors he referred to epidemics as 'pseudo-epidemics' as a means of diminishing their importance, but he never defined how these differed from epidemics proper. Elsewhere, he acknowledged that rates of childbed fever were higher in the winter, but he said this was only because during the hot summer months students preferred to spend idle hours picnicking in the Sun rather than studying in the morgue and immersing their hands in cadaveric poisons. As his readers quickly realized, this was special pleading of a most unconvincing kind.

In view of Semmelweis' failure to explain one of the most distinctive characteristics of outbreaks of childbed fever—their epidemic nature—one can entirely understand the statement of the great chemist, Justus von Liebig, that while the 'perspicacity of Dr Semmelweis ... cannot be doubted' it is 'evident that there are also other significant causes of puerperal [i.e. childbed] fever'. Karl Braun, Semmelweis' successor in Vienna and now a well-respected obstetrician, also admitted that dead, organic matter could produce the infection, but he appended the Hungarian's theory to a list of twenty-nine other potential causes. Braun, who eventually succeeded Johannes Klein as professor of obstetrics, noted in his book *On the Puerperal Processes*, 'we hardly find any authority to support the practical application of the infection theory.' His one concession as professor was to forbid students from taking part in deliveries if their hands actually stank of corpses. The distinguished Würzburg professor of obstetrics, Wilhelm Scanzoni, likewise conceded that 'in certain cases' infections like those cited by Semmelweis 'may occur'. He added, however, that 'the epidemic nature of this evil cannot be denied,' in consequence of which Scanzoni was 'still of the opinion that it is chiefly miasmatic influences in lying-in hospitals which are the root of the disease'. In other words, airborne poisons did most of the damage.

Instead of trying to deal with the points they raised, Semmelweis flung charges of 'murderer' and 'assassin' at these highly respected medical authorities. But their arguments were perfectly legitimate. Liebig, Braun, and Scanzoni were first-class medical scientists who had introduced many innovations of their own: they weren't ideologues but they quite reasonably thought Semmelweis was. As they repeatedly explained, in most maternity hospitals childbed fever disappeared for months at a time only to reappear suddenly and destroy dozens of lives. This just could not be made to fit with Semmelweis' claim that the disease arises due to routine unhygienic practices. If dead, organic particles were really to blame, then all hospitals in which doctors regularly delivered babies after performing autopsies or touching patients with open wounds should never have been free from fever. This, however, was not the case. As the eminent German pathologist, Rudolf Virchow, recalled, 'I had the very best result in the treatment of puerperal fever, although I was daily engaged in dissections, in touch with cadavers and cadaveric parts.' Conversely, many more doctors added, in some wards where expectant mothers were isolated from all sources of 'dead organic' matter, childbed fever nevertheless struck repeatedly and savagely.

Yet an important question remains. Semmelweis' theory might have been dubious, but why wasn't the striking success of his introduction of disinfectant on its own enough to convince obstetricians that he was on to something? Bringing about a 70 per cent decline in childbed fever mortality was a truly exceptional achievement. However, once more the epidemic character of childbed fever muddied the waters. As we've seen, the fact that the infection tended to come and go from most lying-in wards was very difficult to make fit with Semmelweis' explanation for Ward One's mortality statistics. One critic, the obstetrician Eduard Lumpe, put it like this: where there were 'incredible variations in the incidence of sickness and death ... any other possibility is more plausible than one common and constant cause.' It was a cogent observation: the epidemiological picture didn't tally with the claim that all cases were due to the transmission of dead organic matter. And if, as seemed to be the case, childbed fever could be caused in a variety of different ways, then hand-washing alone was unlikely to have produced Vienna's sharp decline in fatalities. Since Skoda's official commission was never convened, this was a hard argument to gainsay. Playing safe, Lumpe himself decided to 'wash and wait,' but he was convinced that Semmelweis had not furnished a complete explanation for the mortality decline in Ward One. So, even those who accepted Semmelweis' statistics as reliable did not consider hand-washing the *only* cause of the drop in mortality.

It was a view that gained considerable strength in October and November 1848, when two women were admitted to Ward 1 with vaginal lesions freely discharging the kinds of 'decomposing animal organic matter' that Semmelweis believed to be the sole cause of childbed fever. As might be expected, he immediately blamed the sudden rise in mortality that ensued on the transfer of matter from these women to others on the ward. But, as Semmelweis' many critics observed, the two women themselves did not develop childbed fever. How, then, could suppurating matter be the primary cause of childbed fever when two women with serious vaginal lesions had failed to succumb to it?

In response to these data, many of Semmelweis' erstwhile supporters cut their losses and returned to his initial 'cadaveric poison' thesis. In doing so they fatally weakened his case, for, as noted earlier, the majority of other lying-in hospitals' midwives and obstetricians had little or no contact with cadavers. Cadaveric poison could not, therefore, be the leading cause of childbed fever. Either way, it was now harder than ever to believe that hand-washing had really made a decisive difference. It seemed that a factor yet to be discovered was actually responsible for the downturn in deaths achieved in Vienna's Ward One.

And then came the germ

Ignaz Semmelweis could explain neither the fever's seasonal patterns of incidence nor its habit of coming and going at various times in the year even though hospital practices had changed not one iota in the intervals between its unwelcome visits. Of course, the missing ingredient that would later account for these anomalies was the modern germ theory of disease. Largely because we take the role of microbes in causing disease for granted, Semmelweis' critics now look like ill-natured and ignorant fools. But it's only with the knowledge of germ theory that his opponents lacked that we can understand the epidemiological features that so bemused mid-nineteenth-century doctors.

It's now obvious that seasonal variations in the incidence of childbed fever occurred because the bacteria responsible bred at different rates depending on ambient temperatures. Again unlike doctors of Semmelweis' period, we can also see that the infection often struck wards where there were no suppurating wounds or cadavers to dissect because deadly germs can be found in many places. Corpses and wounds are rich, but not exclusive, sources of harmful bacteria. Modern germ theory can also explain why Virchow and other surgeons enjoyed low levels of childbed fever despite regularly coming into contact with corpses and the vilest open sores. First, since specific kinds of infection are caused by specific microbes, there was no certainty that any given body or wound would contain the germs capable of generating childbed fever; often they did not and those subsequently touched had a lucky escape. Second, sometimes medical staff did transfer the bacteria responsible for childbed fever, but the particular strains were so mild that the mother's natural defences easily fought back. Knowing no better, their doctors supposed the wards to be fever-free.

In addition, before the advent of germ theory neither Semmelweis nor his critics could possibly understand that without a continuous supply of new hosts all epidemics rapidly die out. In large hospitals with a constant flow of expectant mothers, like Vienna's, childbed fever rarely went away. But the typically much smaller institutions in other parts of Europe and America often entirely escaped the disease for weeks or months at a time. This is because in maternity wards containing just a few women, the bacteria soon ran out of hospitable bodily environments, and infection recurred only with the arrival of new batches of patients who lacked immunity. This variation in the size of hospitals produced a highly complex epidemiological picture that no doctor of the mid-1800s could possibly have understood. As a result, those working in small or medium-sized clinics who read Semmelweis' *Aetiology* hardly recognized the disease he was describing. And, in the absence of germ theory, there was no way in which its author could adjust his thesis in order to accommodate these discrepancies. Likewise, without the sophisticated understanding of immunity that came in the wake of germ theory, obstetricians couldn't see that the two women with the discharging lesions who failed to develop childbed fever had probably acquired an ability to fight off infection while being quite capable of infecting others.

Of course, in an age in which germ theory was little more than a fanciful speculation, Semmelweis could hardly have been expected to get it all right. Equally, however, we cannot condemn Semmelweis' critics for attacking his theory, because in the absence of knowledge of germs Semmelweis could explain only a few features of epidemics of childbed fever. Making the best use of the data and theoretical possibilities open to them, Liebig, Braun, and Scanzoni gave the Hungarian his due and then continued the search for causes that would explain the features of the disease that he elected to ignore. Ultimately, of course, the arrival of modern germ theory made these debates obsolete, but not because Semmelweis was proved to be comprehensively correct. The revolution inaugurated by Louis Pasteur and Robert Koch did not just vindicate Semmelweis; to a large extent, it also confirmed the wisdom of Liebig, Braun, and Scanzoni in attacking him. All the serious players in this debate were both right and wrong; but there can be no doubt that Semmelweis was in the possession of the largest blind spot and was much the most vitriolic in defending his position. This puts the way Semmelweis responded to his many critics in a different light.

Medical Neros?

Semmelweis tended to bottle things up. When he was ignored, slighted, or insulted he bore it with seeming equability, and many of his critics could have been forgiven for thinking that he was unusually resilient or simply too arrogant to care. In fact he was a deeply sensitive individual who absorbed criticism until some kind of inner limit was reached. Then his sense of hurt was decanted in sudden violent outbursts that none could have predicted. This character trait gave Semmelweis an unenviable gift for turning respectful critics into implacable enemies.

As we've seen, Semmelweis only sat down to write his treatise on childbed fever over a decade after his introduction of hand-washing in Vienna. Then he did so at a feverish pace. Hardly ever looking back to what he had already written, sending sheets off to the printers as soon as the ink was dry, and burying his most fundamental ideas beneath a morass of incoherent digressions and frenetic ramblings, he ensured for his 1860 treatise a low and mostly unsympathetic readership. But few of the critics realized just how much Semmelweis laid his ego bare in this book, nor how difficult this acutely sensitive man found it to put pen to paper. Thus, having absorbed the criticisms of Klein, Liebig, Braun, and others, Ignaz Semmelweis could take no more and erupted in righteous indignation. No longer able to see the manifest problems with his theory, he attacked those who opposed him with a scientifically suicidal zeal.

In late 1861 he wrote the first of several open letters to a Viennese professor. Calculating the number of lives he believed could have been saved by hand-washing, Semmelweis signed off with the embarrassingly forthright remark, 'In this massacre, you, Herr Professor, have participated.' Next he addressed Scanzoni. 'Your teaching,' he declared, 'is founded upon the dead bodies of lying-in women murdered through ignorance, and because I have formed an unshakeable resolution to put an end to this murderous work, I call upon you to comply with my request.' He then added, 'You have demonstrated . . . that in a new hospital like yours provided with the most modern furnishings and appliances, a good deal of homicide can be committed, where the required talent exists.' He later recommended to Braun that he 'take some semesters in logic'. And to another doctor he railed, 'I proclaim you before God and the world to be an assassin.' As one Berlin professor complained, 'Semmelweis calls everyone who disagrees with him an ignoramus and a murderer.' All radical scientific ideas require excellent public relations. They also need high-status support. Semmelweis couldn't provide the former and he systematically alienated the latter.

Perhaps his most fatal error was to aggravate Rudolf Virchow. Probably the greatest physiologist of his day, Virchow was incensed when he read the following words written by Semmelweis, 'there are at present 823 of my pupil midwives . . . who are more enlightened than the members of the Berlin Obstetrical Society; should Virchow have given them a lecture on epidemic puerperal fever they would have laughed in derision.' Not too big to take offence, Virchow thereafter used every opportunity to rubbish Semmelweis' ideas. In this unequal battle, the Hungarian came off a good deal worse. Any chance there had been of gradually coaxing Virchow and others round to his point of view was squandered. No one was prepared to listen to someone so emphatically incapable of assuming the detached air of scientific debate. When Semmelweis should have gone into print, in 1847, he kept silent; now, when virtually every word he uttered was damaging his cause, he couldn't stop writing.

By this stage even Semmelweis' most faithful allies from the Viennese school had also deserted him. Supporting this erratic, irascible Hungarian was doing their own chances of success in the medical world few favours. Yet they too were left smarting for personal reasons. As we've seen, having lost his battle against Klein, Semmelweis quitted Vienna. But he made a foolish error in leaving the city without saying even the most perfunctory of goodbyes to his friends. They found this rudeness hard to forgive. As one historian wrote of Josef Skoda, 'he was deeply hurt by the ingratitude and folly of Semmelweis. He said nothing, but for him Semmelweis ceased henceforth to exist.' No longer were Semmelweis' comrades willing to stake their reputations on the ideas of a man singularly bereft of good judgement. Although it's a pity that in science the personality of its practitioners makes such a difference, if we examine our personal affairs, few of us will be surprised. In any case, Skoda, Hebra, Rokitansky, and the others had their own interests and were quite busy enough making important advances in other medical fields to take up the colours of a less than gracious ex-colleague.

The real severity of Semmelweis' insanity sheds further light on why he managed to retain so few allies. As one historian has noted, his 1860 treatise betrays signs of 'mental aberration and feelings of persecution' and, according to one close friend, Semmelweis had already begun defending 'his scientific views with a passion bordering on fanaticism'. In early 1865, however, he became thoroughly unhinged. This hitherto respectable and temperate man took to heavy drinking. He no longer cared about his family, and he became obsessed with achieving sexual gratification. Spurning his wife, his sexual appetite was, according to the same friend, routinely satisfied through, 'intercourse with prostitutes . . . frequent masturbation, and indecency towards both acquaintances and strangers'. Whether he'd experienced a psychotic break because of the strain of his ideas being so unsympathetically received, or whether he suffered from an organic condition such as neurosyphilis, will never be known with certainty.

What we can say is that Semmelweis became a serious embarrassment both to his respectable mercantile in-laws and to his professional colleagues. Papers released by the Austrian government during the 1970s now reveal that he had become so insane that his family and erstwhile friends were prepared to go to extreme lengths to have him incarcerated. In the spring of 1865, a plot was hatched in which his wife's family and several of his medical colleagues all seem to have been complicit. On 29 July, encouraged by his wife, he climbed aboard a train he thought would take him to a German spa for some much-needed relaxation. When it stopped in Vienna, he was calmly led away by his uncle-in-law and an old friend from the Vienna hospital. With the utmost stealth, they took him to a low-grade asylum, managed to bypass the usual admission procedures, and then quietly left. Too late, Semmelweis realized what had happened. He struggled, lashed out at the guards, received a savage beating, and was hurled into a dark cell. When his wife tried to visit the following day, perhaps to express her horror at the subterfuge carried out by her family, she was turned away.

Thirteen days later his corpse said it all. A gangrenous finger, almost certainly arising from an infected cut sustained in a fight with the warders, was the actual cause of death. But severe bruising and internal injuries accompanied his rotten finger, plus a profusion of boils on his extremities caused by the filthy conditions of the cell. That this highly educated man was not taken to a private clinic; the conspiracy of silence that ensued; the wanton brutality of the warders; and the fact that Semmelweis was ill-treated despite his social standing, strongly suggests that his death was welcomed, if not prearranged. Once again, this is a matter never likely to be fully resolved.

All we can deduce with confidence is that Semmelweis was sent to an asylum much beneath the dignity of a member of the educated classes and one where the staff were a lot more venal than those in more respectable institutions. In such a place there would be no formal enquiries or diagnoses and certainly no voluntary discharges. Even if he survived the warders' blows, Semmelweis' in-laws and medical colleagues could rest easy that he would be unlikely to cause them further embarrassment. Semmelweis had become such a liability that both family and friends could see no alternative but to have him locked away in perpetuity. Considering the extent of his mental disorder, it's little wonder that his fellow obstetricians ignored him during his last years of life and preferred to forget about him after his death.

The spin doctors

At first glance Semmelweis seemed to be the perfect hero, but on closer examination a much more complex picture has emerged. And the final question we have to ask is how, given both his own and his theory's failings, Semmelweis managed to achieve such an extraordinary level of posthumous fame.

By the late 1880s, with the germ revolution well underway, it was very hard for doctors to understand why he'd been so widely repudiated. Unable to see the world as it had looked before the advent of germ theory, the hostility of Liebig, Braun, and Scanzoni made sense only as an ignoble expression of overweening pride.

Furthermore, at a time when the germ theorists' greatest triumphs stemmed from the introduction of strict hygiene measures in hospitals and operating theatres, they could hardly be unmoved by the relentless campaign Semmelweis had waged to have bowls of chlorinated water placed in maternity wards. He was a hygienist before his time and, according to the fuzzy logic of romantic history, this made him even more accomplished than those who later did the same on the basis of better theory and more cogent data. Yet there's more to Semmelweis' posthumous rehabilitation than a lack of contextualization.

In 1892, a handful of Britain's leading medical authorities gathered at the London residence of a senior surgeon. Their brief was to devise the 'best way of promoting a memorial of Semmelweis and his great services to humanity'. So unbounded was the adoration they expressed for him that the gathering might have been mistaken for a prayer meeting. After several further conferences, held over the course of more than a decade, a statue was finally erected in Budapest in the presence of his widow. But there was a certain irony in this willingness of British doctors to pay such generous homage. As one obstetrician pointed out, the Hungarian had been singularly rude about British midwifery practices throughout his career. More significantly, the 'rediscovery' of this pioneer of hospital hygiene caused some embarrassment for the pioneer of antiseptic surgery, Baron Joseph Lister. Rumours quickly spread that Lister had simply copied the Hungarian's ideas, and he was obliged to make a statement to the press declaring, 'Semmelweis had no influence on my work.'

Nevertheless, it's not hard to understand why Joseph Lister should have been happy to underwrite the canonization of Semmelweis. Never shy in promoting himself, Lister must have seen that he could enhance his own reputation by publicizing a story that implied that he too had taken a courageous risk by challenging the opponents of antiseptic practice. Furthermore, how many self-made men would cavil if they discovered in later life that they were the descendants of a noble family? Lister, a pioneer of clean surgery, found out that he was nowhere near as original as he'd supposed. But any disappointment would have been short-lived, for so exquisitely did Semmelweis seem to fit the template of the romantic hero of science that celebrating his achievements could only improve Lister's standing by adorning the profession to which he belonged. Where medicine had so often been associated in the public mind with the pompous and ineffectual physician, the quack's dangerous tonics, and the incalculable agony of the operating theatre, this obscure Hungarian's other-centred pursuit of the truth promised to bathe the progressive wing of the medical community in an almost ethereal light. The legend of Semmelweis gave them a founding father of truly epic dimensions.

Thanks to Louis Pasteur, Robert Koch, Joseph Lister, Florence Nightingale, and others, by the 1890s medicine could boast a string of successes. Eminent doctors and surgeons were receiving state funerals. Streets, squares, and parks were being named after them, and the statues in European cities now celebrated doctors as well as generals, politicians, and heads of state. Perhaps because so many had begun to question their

The Semmelweis memorial in Budapest (statue by Alajos Stróbl), unveiled 30 September 1906.

Christian faith, the public had a hunger for stories of scientific courage and self-sacrifice. And the medical profession happily exploited this fondness for hero worship in order to raise its own standing. But for all its triumphs, medicine still didn't have a heroic drama that stood any kind of comparison with Galileo's persecution by the Roman Catholic Church, the martyrdom of the Oxford clerics, the charge of the Light Brigade, or the death of Gordon of Khartoum. Lister may have served the hero's obligatory years in the wilderness, but he'd since received compensation with interest. And although doctors could boast of the intellectual prowess of Pasteur and Koch, both men's life-stories lacked the pathos required for a truly romantic hagiography.

This is where Semmelweis fitted in. Here was the ultimate lone, embattled genius, the man before his time who had the insight and the courage to fight for the truth even though, as the historian F. H. Garrison lamented, it 'brought on insanity and death'. It was simply wonderful copy for this newly confident, assertive, and self-promoting medical fraternity. Semmelweis had lived and died in selfless devotion to child-bearing women, and his sacrifice lent to the banner of medical science the rich hues of nobility and chivalry. Thus, smoothing over their subject's flaws and vilifying his opponents, the hagiographers created perhaps the finest tragedy in the history of modern medicine.

Yet the true story, painstakingly pieced together in recent years, reveals a life of even more intense tragedy. For it now seems clear that had Semmelweis chosen to deal calmly with the entirely legitimate objections so many obstetricians had to his theory, he might have avoided a miserable death. More disturbingly, had he been less vitriolic in attacking his critics he might even have persuaded some of them that routine hand-washing was at least worth a try.

But, all his sins forgotten, in September 1906 a statue of Semmelweis was unveiled in Budapest. Gathered together at the grand ceremony were his widow, Maria, as well as several ex-colleagues and surviving members of his family. Given that these guests were only too aware of Semmelweis' flaws of character, his infelicities, and the unhelpful vehemence with which he'd attacked his opponents, one wonders how they felt as they took part in the unveiling ceremony, surrounded by some of the world's most accomplished doctors and scientists. Perhaps some of them were secretly sensitive to parallels that could be drawn between his sparsely attended funeral in August 1865 and the lavish celebrations four decades later. But whatever their inner feelings might have been, they performed their assigned roles to perfection.

PART THREE

HEROES MADE TO MEASURE

HEROES MADE TO MEASURE

> Today no one bestrides our narrow world like a colossus; we have no giants.
>
> Arthur M. Schlesinger, *The Coming of the New Deal* (1958).

Modern society, it is often claimed, has too few heroes. For every Mother Theresa or Martin Luther King, the present age has thrown up a myriad celebrities with neither the sense of duty nor the capacity for courageous self-sacrifice that once marked out the true hero. In the political sphere, we no longer seem to produce individuals of the calibre of, say, a Thomas More or an Abraham Lincoln. In the world of medicine, where are our Kochs, Pasteurs, or Nightingales? Among social reformers, who could we now mention in the same breath as Dorothea Dix or Elizabeth Fry? It might be said that we're no longer interested in heroes. But the popularity of films with as varied a connection to reality as *Apollo 13, Braveheart, Star Wars*, and *Lord of the Rings* shows an enduring public appetite for the outstandingly noble. The hugely successful series *Star Trek* is, moreover, all about good beings doing good things in what can be a very naughty universe.

It's not the human desire for heroes that's wanting, so perhaps the dearth of real-life heroes has more to do with a change in what the media considers worth reporting. Instead of presenting outstanding individuals to be revered and emulated, modern media appear to be much more inclined to offer us the lives of the merely famous, people with some measure of talent but seldom of the timeless variety. As the critic and historian, Daniel Boorstin, wrote in his 1961 polemic, *The Image*, there seems to be nothing more newsworthy today than transient celebrities doing perfectly ordinary things, be it shopping, sunbathing, getting drunk, or falling out with their lovers. And beneath the talented, there now lies a raft of people bent on fulfilling Andy Warhol's prediction that we'll all enjoy our fifteen minutes of fame. These are the cannon-fodder of reality TV shows, the foot soldiers in the press's insatiable drive for human-interest stories, individuals so sold on the idea of fame that they will do anything to secure it.

Perhaps the biggest indictment of the new relationship between press and public is what is known as 'the tall poppy syndrome', the media's taste for first building up an individual and then tearing them down once they put a foot wrong. Readers inveigh against inappropriate behaviour, then reward it by buying extra copies. The fall from grace of the British politician, John Profumo, in the early 1960s, following his affair with Christine Keeler, must have consumed whole forests of wood pulp; the forty years since then that he has spent working selflessly for an inner-city charity gets but an occasional mention.

But the media are not solely responsible for these trends. Media barons assure us that they're merely giving the public what it wants, and it's a claim with at least some validity. How is it, then, that in earlier periods public figures were represented so much more generously? Certainly, the press was once comparatively deferential. During Franklin D. Roosevelt's presidency, most journalists consented to his wish to keep pictures of him in his wheelchair out of the paper. Many colluded with the staging of photographs showing him standing erect. Nor did earlier generations of journalists pass much comment on Lloyd George's sexual appetite or Churchill's taste for strong liquor. Needless to say, one can hardly imagine their modern equivalents behaving with such restraint.

To many of us this change in tone of media coverage is to be regretted. After all, people of exceptional talent are not so common that they can be easily replaced once broken at the whim of a latter-day William Randolph Hearst. So should we look back wistfully upon these gentler times? My own feeling is that we would be unwise to do so. While elements of the modern media may have over-corrected their product, spurning decency alongside deference, the fact is that few of those celebrated by our ancestors for their integrity and selfless public spirit actually lived up to their near-perfect public images.

It rarely takes much archival digging to see through the mystique of famous statesmen, reformers, and military leaders. Not all, of course, were venal and corrupt, but some undoubtedly were. The Victorian writer, Thomas Carlyle, though convinced that national greatness depended on heroes and hero-worship, perceptively remarked, 'no man can be a hero to his valet.' He was surely correct. With an intrusive and undeferential press, combined with the greater accessibility of state documents, we know too much about the tawdrier side of the exercise of power to be avid hero-worshippers. But we can at least rest assured that we're far less commonly duped than our predecessors were. Losing the imagery of the pristine hero is the price we pay for accountability.

Misrepresentation in science is typically less worrisome than in politics, not least because in science power is seldom concentrated in just a few hands. But, as the chapters on Lind and Semmelweis indicate, the lives of individual scientists are also often embellished and made to fit crudely heroic caricatures of selfless brilliance. Rarely are these distortions effected with a deliberate intention to mislead. Heroes of science are more often constructed in subconscious fulfilment of the standard model of scientific discovery. There are, however, several instances of myth-makers having resorted to more serious manipulations underpinned by deliberate falsehoods. This section of the book examines two such cases.

Johann Weyer and Philippe Pinel have both been dubbed by different writers as the 'true fathers' of modern clinical psychiatry. Weyer is celebrated for being the first to deny that the insane, including those who willingly confessed to witchcraft, were subject to demonic possession. He proposed, it's claimed, that accused witches receive humane medical care, not bloody persecution at the hands of the Inquisition. Pinel is remembered for an event supposed to have taken place in the immediate aftermath of the French Revolution: the unshackling of hundreds of terrified and brutalized patients from two of Paris' largest 'madhouses'. This bold gesture is seen by many as marking the advent of the enlightened medical care of the insane, a signal victory of humanism over barbarism, and the realization of Weyer's sixteenth-century dream.

In reality, as we'll see shortly, Weyer never for a moment doubted that most of those who confessed to witchcraft actually were possessed by Satan, and Pinel played only the most trivial of roles in the unchaining of France's insane. But the most interesting aspect of each story is how the myth arose and then achieved its ascendancy. In both cases the insecure status of psychiatry as a medical speciality in France during the 1800s was a key factor. A nascent and embattled discipline, it struggled to discredit theological definitions of insanity and thus usurp the authority of the clerics and laymen who had traditionally cared for the mentally ill. As part of a general campaign to eject competitors from their patch, leading psychiatrists systematically reconstructed the stories of Weyer and Pinel. The mythical heroes they created bore only an oblique resemblance to the actual historical figures. Perhaps, as Sellars and Yeatman, the authors of the satirical *1066 and All That*, would have said, it was 'a good thing' for us that the psychiatrists' campaign was so successful. None the less, both myths remind us that science, like politics, has its share of propagandists.

VINCE TEIPSVM.

EFFIGIES IOANNIS WIERI ANNO
ÆTATIS LX. SALVTIS M.D.LXXVI.

Johann Weyer (1515–1588)

> The fervent, revolutionary humanism of Weyer
> stands out as a phenomenon which must be assessed
> not merely as a striking episode in medical history,
> but as a momentous step in the whole history of man.
>
> Gregory Zilboorg, *The Medical Man and
> the Witch during the Renaissance* (1969).

In 1486 two Dominican friars, Jacob Sprengler and Heinrich Kramer, wrote a book that in the long run did very little for the reputation of their order. Entitled *Malleus Maleficarum*, but usually referred to as *The Witches' Hammer*, it's among the most gruesome manuals ever compiled. Sprengler and Kramer's aim was to provide officers of the Holy Catholic Church with a handbook that would enable them to identify, convict, and punish witches. But *Malleus Maleficarum* was more than just a do-it-yourself guide for witch-finders. It was also a strident appeal to good Christians to mount a new crusade. This time it was to be directed not against the Muslim infidel but the enemy within: a Satanic fifth column thought to be leading God's chosen peoples into all manner of ungodly sins.

Having dubbed themselves the 'dogs of the Lord' (*canes Domini*), Sprengler and Kramer carefully set out the 'proper' means of eradicating witchcraft. Condoning all conceivable forms of torture, they insisted that accused witches must always be brought to confess. A confession having been obtained, only one sentence was then permissible: death by burning as a foretaste of Hell. Suffused with this kind of brutality, *Malleus Maleficarum* is said to have unleashed upon Europe wave after wave of murderous hysteria. Witch trials had been a rare event before the fifteenth century, and burnings rarer still. Now, at least in part due to Sprengler and Kramer, inquisitors visited nearly every part of the continent. For those convicted of witchcraft, there was no possibility of returning

LEFT: *Portrait of Johann Weyer, a woodcut of 1577.*

apologetically to the fold. According to our two friars, all witches willingly offered up their souls to the Devil and could therefore be tried as both heretics and criminals. On this basis, thousands of the accused were killed in a variety of horrifying ways.

Amid this frenzy of inhumanity one brave man spoke out. His name was Johann Weyer. The second son of a wholesale trader in hops, Weyer had risen to become physician to Duke William III of Berg, Julich, and Cleves, lands situated on the lower Rhine. While the rest of his profession either ignored or colluded in the brutal campaign of witch-burning that stretched from the late 1400s to the mid-1600s, he felt compelled to protest. He did so in the form of a book published in 1563, entitled *De Praestigiis Daemonum*. Accusing the witch-finders of barbarism was always likely to get Weyer into trouble, and it did. Even warring Catholics and Protestants were united in their condemnation. Johann Weyer's book was burnt by the Lutheran University of Marburg and put on the Index of Forbidden Books by the Catholic governor of the Netherlands, the Duke of Alba.

Luckily, a few copies of *De Praestigiis Daemonum* survived. And, when it was reread in the 1800s, these provided a basis for the recasting of Weyer as a quintessential man before his time. But it wasn't only for his humanitarianism that Weyer was acclaimed. More important, he was posthumously credited with being no less than the 'father of psychiatry'. A series of admiring portraits told how, risking the wrath of a vengeful and bloodthirsty Inquisition, he attempted to suborn superstition in favour of humane science. With stunning prescience, it was said, he argued that those who confessed to witchcraft were not possessed, but guiltless victims of mental illness deserving not persecution but the enlightened care of physicians. His heroic stand against the witch-finders thereby marked the official birth of the scientific approach to mental malady.

This triumphalist reading quickly became the accepted view. Sigmund Freud identified *De Praestigiis Daemonum* as being among the ten most important books ever written. By the time Gregory Zilboorg and George Henry came to write their classic, *A History of Medical Psychiatry*, in 1941, Weyer's status as father of the discipline was unassailable. The clearest formulation of the legend came in 1969 with the publication of Zilboorg's

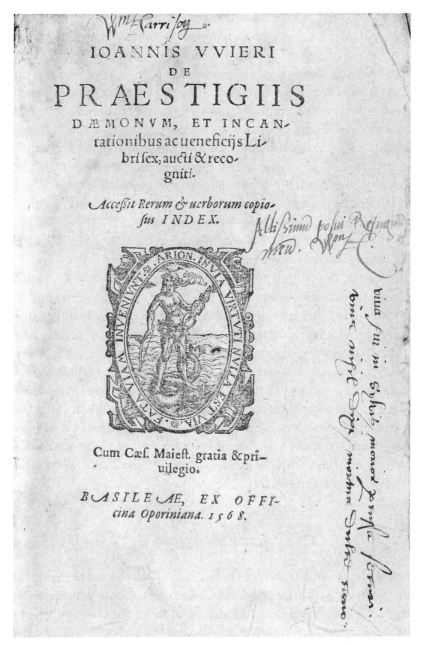

IOANNIS VVIERI

DE

PRAESTIGIIS

DÆMONVM, ET INCAN-
tationibus ac ueneficijs Li-
bri sex, aucti & reco-
gniti.

Accessit Rerum & uerborum copio-
sus INDEX.

Cum Cæf. Maieft. gratia & pri-
uilegio.

BASILEAE, EX OFFI-
cina Oporiniana. 1568.

The frontispiece of Johann Weyer's De Praestigiis Daemonum, *published in Basle in 1568.*

The Medical Man and the Witch during the Renaissance. He eulogized the way that Weyer's psychiatric approach to understanding witchcraft confessions 'instigated a revolution in medical thought'. But Zilboorg explained that this was no overnight revolution. *De Praestigiis* was so far in advance of its time that it laid dormant for two and a half centuries. Then medical men of the early 1800s read Weyer's great book and, inspired by its benevolent vision, proceeded to establish psychiatry as a professional medical discipline based around humanely managed asylums. This heroic characterization of Johann Weyer persists to this day. The *History of Psychology and Psychiatry* devotes a chapter to him, entitled 'The Father of Modern Psychiatry'. The *Comprehensive Textbook of Psychiatry* includes a long section, 'Johann Weyer: A Pioneer of Modern Psychiatry'. And in Hugh Trevor-Roper's *The European Witch-craze* he is represented as the humane foil to the vicious agents of the witch-hunt.

For all this enthusiasm, except among those with a detailed knowledge of the history of psychiatry, Weyer remains a nonentity. So the reason for telling his story isn't to shatter a popular myth. Instead, the significance of this case lies in its showing from what unpromising materials a heroic reputation can be forged where there's the will and the imagination to do so. Weyer was in every sense a man of his time and his place. But to psychiatrists in the 1800s it was more convenient to represent him as a true scientist out of place in an age of superstitious barbarism than it was to provide an accurate portrait of this courageous but in other respects unexceptional man. Why they felt the need to deceive themselves and others in this way will be explored once we've shown just how fatuous the legend of Johann Weyer is.

Crazy old women

Few myths are entirely the work of the imagination. Behind the legend of the mermaid, for instance, lies the manatee: grotesque in relation to the myth, but real nonetheless. To some extent this is the case with Johann Weyer. For, trawled with the right-sized net, *De Praestigiis* will give the reader a fair haul of seemingly forward-looking attitudes.

Weyer is said to have adopted a purely naturalistic view of witchcraft, arguing that apparent cases of devilry could easily be explained in terms of the operation of natural laws. And, consistent with this claim, he did indeed attack witch-finders for reading magic and miracles into even the most benign natural phenomena. 'Many things come before our eyes from time to time which are thought to be beyond the law of nature,' he wrote, 'and deemed to be the mocking activities of demons, even though Nature—the parent of all things—has produced them from definite causes not difficult to understand.' Weyer cited several examples of superstition getting the better of the witch-finders. There was, for instance, the *lunaria* plant (popularly known as 'honesty'). Its flattened, translucent seed heads were widely believed to be 'apparitions of evil import'. But students of nature, Weyer noted, knew perfectly well that *lunaria*'s glow was simply a natural property of the plant itself.

More important, from our point of view, were Weyer's remarks on the states of mind of those who confessed to being witches. Here too one can find remarks that make him sound like a modern in sixteenth-century clothing. First, Weyer implicated mental illness in witchcraft confessions. Old women who confess to having killed neighbours, ruined crops, and taken part in orgiastic sabbats are actually suffering from weakened wits, he asserted: their perceptions are 'dulled by age, or inconstant by reason of her sex, or unsteady because of . . . weak-mindedness, or in despair because of a disease of the mind'. This sickness of the mind, Weyer argued, could take various forms, but anyone who claimed to have superhuman powers was in a delusional state. Under no circumstances were their lunatic ravings to be believed. Second, Weyer identified a physical cause for the kind of delusions that convinced old women that they could perform diabolical feats. In most cases, he argued, an excess of black bile (a bodily humour believed to be secreted by the kidneys and spleen) had somehow fuddled their intellect. Third, Weyer stipulated that in cases of suspected witchcraft the physician ought to be called in before the priest. 'On this matter,' he wrote, 'let the judgement of the physicians be consulted— physicians renowned for their understanding of natural objects . . . just as the law wished in other cases which fall in the medical sphere.'

All this has a modern ring to it. Weyer seemed to be claiming that witches were entirely innocent of consorting with the Devil, that they confessed to outlandish crimes as a consequence of madness, and that they ought to have recourse to physicians who could address the physical causes of their complaints. Some of Weyer's modern devotees might shift uncomfortably at his commitment to the ancient medical doctrine of the four humours, but in a rare show of historical sensitivity, most of his admirers have declared that this is unimportant. What really matters, in Zilboorg's words, is that 'he knew that the answer to the problem was to be found in medicine.' Weyer, then, appears to have been deeply sceptical about the reality of witchcraft in a period awash with ignorance and superstition. Led by a coolly rationalist outlook, he perceived the folly of the witch-finders and sought to put a stop to their superstitious atrocities. But how accurate an impression is this of Weyer's attitudes to witchcraft?

'The Devil's bath'

Reading the standard historical accounts of Johann Weyer and then looking closely at *De Praestigiis* is akin to arriving in a foreign country, having read a guidebook that gets a few trivial details right and everything else, including the name of the capital and the local language, totally wrong. It's a disorientating experience, for the astonishing fact is that Weyer never did argue that witches were suffering from insanity alone. In nearly every case of suspected witchcraft, he believed that the Devil had indeed inveigled himself into the soul of a crazy old woman. For Weyer, as for Robert Boyle and Joseph Glanvill, witches unquestionably existed.

Weyer did think physical malady and old age to be part of the overall picture, but this was anything but original. Like most contemporary physicians, Weyer alloyed his naturalistic theories with a spiritual element. In the case of witchcraft, he argued that the Devil is drawn irresistibly to those whose bodies are choked with black bile, inducing melancholy. The 'crafty spirit', Weyer wrote, is attracted to 'the sooty vapour of black bile', and 'takes great delight in immersing himself in this humour as being the proper moisture for himself'. Following St Jerome and Martin Luther, plus a host of less well-known theologians and physicians, Weyer referred to black bile as 'the Devil's bath'. To Weyer the natural and the supernatural operated together. The Devil worked his mischief by stirring up the black bile in the region of his victims' 'optic nerves' and thereby tricked them into thinking that they had magical powers. 'With its assistance,' he wrote, 'he induces wondrous phantasms and rare imaginings.' The hapless witch 'sees' things that aren't actually happening, 'so that these deluded women dare to affirm under threat of death that they have seen or done things which have never been seen and have never existed in reality'. Weyer's Devil was a malicious trickster who cheated his victims into making absurd confessions or, less often, committing heinous acts. Old women who confessed to outrageous crimes were delusional, but their madness arose from the malicious offices of Satan himself.

Weyer didn't even consider physicians competent to treat cases of demonic possession unaided. Although he believed that purging black bile might sometimes work and was worth trying in most cases, he stressed that the 'prudent and circumspect physician' interfered as little as possible when spirits were involved. 'When he sees that the evil surpasses natural limits,' Weyer wrote, 'and when he detects the movements and activities of Satan, who is a spirit, he will transfer the burden of cure to a spiritual physician—a clergyman or blameless minister of the Church.' The clergyman must then 'gird on spiritual armour' and confront the Evil One with his faith in the superiority of God. The best approach, according to Weyer, was a combined operation in which the physician first removed the black bile and rendered the victim's body a less congenial home to resident demons. The clergyman then followed up with a series of religious incantations that struck directly at the demons themselves. Science and religion were complementary areas of expertise, most effective when working hand in hand.

So far is doesn't seem that *De Praestigiis* could have been of much comfort to the accused witch. But Weyer's genuine humanity is more apparent in his view on the culpability of the witch. Though he didn't deny that witches were in league with Satan, Weyer did assert that they weren't to blame for it. Kramer and Sprengler had urged that witches deserved burning because they willingly entered into alliances with the Evil One. Weyer replied that it was ridiculous to believe that frail, lonely old women, whose bodies coursed with melancholy humour, could resist the blandishments of a crafty and powerful Devil. 'The singular debility of their age or simpleness of their sex should,' he wrote, 'exonerate them or at least mitigate their punishment.' Witches were innocent parties. Able to put up only the feeblest resistance, they were no more than the Devil's dupes.

In an attempt to shore up this defence, Weyer added that witches were in any case virtually harmless. Dr Faustus was said to have struck a bargain with the Devil in order to acquire supernatural powers. Weyer denied that any such exchange was possible. The Devil's activities, he argued, were limited by the mental and bodily capabilities of the individuals he possessed. Any attempt on the part of a feeble old woman to invoke the power of the Devil was, he remarked, 'something ridiculous to behold'. And any claims by these women to having magical powers were entirely imagined. Witches deserved to be pitied not punished, and, above all, protected from the misdirected enthusiasm of witch-hunters. We must, Weyer implored, stop the 'constant shipwreck of souls'.

This is a very different Johann Weyer from the one that has now become ossified in psychiatry textbooks. But Weyer went much further than simply believing in witches.

Satan's little helpers

Weyer also believed that Christendom was under siege by thousands of far more numerous and pernicious 'magicians'. We should be especially vigilant, he wrote, in smoking out magicians, men who draw on dark magic to 'surpass the laws of nature'. In the amorphous category of magician he placed anyone who employed spells to create illusions, used magical incantations to read the future, performed rites to raise the spirits of the dead, or ceremonies with the least resemblance to pagan rituals. Weyer also singled out those who trapped demons into rings of 'hard, unbreakable crystal and glass' from inside which the demons had no choice but to do their magician's bidding. Weyer thought that magicians differed from witches in that they consciously decided to ally themselves with the Evil One; led by cupidity or blind ambition, they sought out the Devil in the hope that he would enhance their own powers of deception.

If *De Praestigiis* has a consistent theme it is simply this: the Devil is everywhere. Instead of refuting the possibility of witchcraft and devilry,

Weyer argued that Satan's cunning had been *underestimated*. The first part of the book was devoted to showing just how necessary it was to be aware of the Devil's malicious tricks, and Weyer began by repudiating those who denied the Devil's terrible power. 'I totally reject,' he wrote, 'all [who] contend that demons do not exist in reality.' For Weyer, it was absurd to believe in the literal truth of the Bible (as he did) and yet deny the awesome power of the Prince of Darkness. After all, he argued, it was Satan's temptation of Eve that resulted in the Fall, and his success in corrupting her descendants that had convinced God to unleash the Flood. Everything God created, it seemed, Satan had warped or destroyed.

Nor did Weyer think Satan's power confined to the biblical age. So convinced was he of the Devil's malice that he accepted as unquestionably true the kind of tales many of his contemporaries already dismissed as fanciful nonsense. In *De Praestigiis* we are told of a town in which copulating couples were visited by Satan and stuck together by such potent magic that they could never be separated. Much to the amusement of their neighbours, Weyer records, even the employment of levers in an attempt to prise them apart could not undo the work of the Devil. With equal credulity, he insisted that the Pied Piper, said to have lured away and killed the children of Hamelin, had not only existed but was the Devil incarnate. As far as we can gauge, by the mid-1500s the story of the Pied Piper of Hamelin was already retreating into the realms of folklore; but to Weyer it was the unalloyed truth, confirmed by various sacred books held in Hamelin, which he said he'd personally examined.

Later on in *De Praestigiis* Weyer described a 16-year-old girl he had met and questioned, who claimed to be the victim of a witch's malice. The girl told Weyer that she repeatedly coughed up pieces of rag and old nails from her stomach, presumably put there by witchcraft. At first, Weyer's modernist credentials seem confirmed. He dismissed the girl's claim by showing that a black rag held on the girl's tongue was free from stomach juices. Yet far from accusing her of trickery, Weyer asserted that the Devil himself had made the rag appear in her mouth. He didn't have the power to insert objects inside the stomach; but he'd placed a rag upon the girl's tongue in such a way as to create the impression that she had vomited it up. His intention, Weyer tortuously reasoned, was thereby to make the girl think that she was the victim of witchcraft so that she would accuse a perfectly innocent old woman of a crime perpetrated by the Devil himself.

This case nicely illustrates how Weyer believed the Devil usually operated within the world. Working entirely alone, unencumbered by the puny frames of human hosts, he wreathed his victims in webs of illusion. Satan, 'who is constrained by the will or command of none but God,' he wrote, 'needs the help of no second creature in displaying his power and declaring his actions.' Weyer went on to describe him as a 'crafty old weaver' whose extensive repertoire included being able to 'confound the organs of sight, blind the eyes, substitute false things for true . . . cover over things which really exist . . . and show forth things which in reality do not exist.' Men and women, Weyer warned, were too susceptible to Satan's trickery to be able to trust their senses.

And witches' accusers, according to Weyer, were especially vulnerable. The Devil took particular delight in targeting these arrogant zealots by convincing them that harmless old women were actually consorting with Satan. As a result they became his unwitting agents in providing 'drawn sword and kindling for savage executioners'. Do not allow Satan to hijack the law, Weyer pleaded; only God's reason is immune to his deceits so only He should be allowed to judge the accused. Those who falsely dragged innocent people before the courts and then had them murdered for nonexistent crimes, Weyer fumed, 'greatly assist their champion Beelzebub'. Clearly, this was not an observation that was going to please the witch-hunters. Nor did Weyer's suggestion that even magicians could actually do little harm. Although they supped with the Devil, Weyer explained, he gave the magicians no more than the crumbs from his table. Even those who consulted spell-books ('wagonloads of blasphemy') produced nothing 'truly substantial' beyond a few harmless visions. Weyer thus felt that even magicians deserved only mild punishment, and plenty of help and guidance to bring them back into the Christian fold. Weyer, then, was no less convinced of the reality of witches than the witch-hunters. The only difference lay in his much more compassionate nature.

Whose smokescreen?

For those who still wish to see Weyer as a fully paid-up modern there are a couple of ways in which one might try to play down his comments on the power of the Devil. The first is to claim that all references to the occult in *De Praestigiis* were simply a smokescreen behind which he sought to rescue the victims of the witch-hunts. According to this thesis, he did not really believe in the Devil at all. Instead, he sailed as close to the atheistic wind as prudence permitted, importing superstitious ideas only so as to escape the attention of a monstrous Inquisition. What did it matter that he pretended to accept the existence of Devils so long as he strove to keep vulnerable old people away from the flames?

The second line of defence is that adopted by historians such as Zilboorg and Henry. They portrayed Weyer as a man in intellectual limbo, so far in advance of his time that he lacked the words to articulate properly what his intuition was telling him. When Weyer spoke of 'evil spirits', they assert, he meant 'illnesses', that for him the term 'Devil' was merely a way of referring to something gone awry, and that although Weyer couldn't quite imagine the rise of a mental-health profession, he nevertheless 'felt that psychiatry must be created'. Weyer, in short, was born out of his time into the sixteenth century, straining with all his power and intellect to turn a gut feeling into a revolutionary medical vision.

Neither of these attempts to preserve Johann Weyer's status as the father of psychiatry can account for what we know about him. *De Praestigiis* does not contain the kind of middle-of-the-road treatment of Satan that might be expected of a writer simply trying to avoid annoying the authorities. Weyer, as we've seen, paid a lot more than lip service to the idea that the Devil roams the world. In fact, it would be hard to find a sixteenth-century writer who made more extreme claims about the Devil's persuasiveness and cunning than did Weyer himself.

Nor would he have had much reason to fear for his life had he written a more ardently naturalistic tract. As the witch craze reached its zenith around the year 1600, dozens of writers tried to call a halt by suggesting that witches were figments of the inquisitors' imaginations. None of these individuals, so far as we can tell, were themselves victimized by the witch-finders. The Englishman, Reginald Scot, roundly denied the possibility of witchcraft in his 1584 book, *The Discoverie of Witchcraft*. King James I fumed, but Scot was left unharmed. The French essayist, Michel de Montaigne, also drew little fire for an essay in which he remarked, 'It is putting a very high price on one's conjectures to have a man roasted alive because of them.' Nor was the Spanish inquisitor, Alonzo Salazar y Frias, reprimanded when he wrote to his superiors after a witch panic in the Basque territories in 1611:

> Damage to crops is sometimes the way God punishes us for our
> sins, and sometimes it is a natural consequence of bad weather
> . . . These things occur everywhere whether there are witches
> present or not, and it is therefore most undesirable for people to
> believe that the witches are always to blame.

Far from stressing the naturalistic and playing down the spiritual, Weyer's argument was thoroughly and unashamedly religious. As he himself wrote in a letter to his princely benefactor, 'My object is chiefly of a theological nature: to set forth the artfulness of Satan according to Biblical authority.' This sentence, in particular, serves to nail the notion that Weyer was an unabashed modern. It also provides the key to understanding why a compassionate man like Weyer could work so hard to demonstrate the ubiquity of the Devil.

Paradox resolved

Weyer's outlook on the world was a composite of three main influences. First was the mystic, physician, and philosopher, Heinrich Cornelius Agrippa, whose household Weyer entered at the impressionable age of 15. He introduced the young Weyer to alchemical writings suffused with discussions of magic, demons, and occult forces. Agrippa also set Weyer an example by saving at least one elderly woman from burning by insisting that she was merely suffering from senility. But Weyer's tolerant attitudes also reflected the influence of one of the most brilliant and pacific minds of the age: Erasmus of Rotterdam. Weyer's sincere concern for the welfare of witches and wizards bore a heavy Erasmian imprint.

Erasmus, however, had a generally positive view of mankind's capacity for redemption, which Weyer did not share. And here we see the impact of another of the great revolutionary thinkers of the age: Martin Luther. Even though Weyer was in the service of a Catholic prince, he took his Lutheranism seriously. In particular, he followed the Lutherans in believing that mankind is weak, laden with sin, and unworthy of God's grace. What the Elizabethan poet, Fulke Greville, called the 'wearisome condition of humanity' was built into Protestantism in two ways. First, Luther argued that we're all born riddled with sin, stained indelibly as the result of Eve's transgression in the Garden of Eden. Second, for the Protestant, salvation could only be won through a genuine love of God. This happy state required that individuals first render themselves worthy of His love, no small task for a being as depraved as man. How could it be achieved? Luther insisted that to find spiritual peace Christians had to dig deep into their souls, acknowledge their unworthiness, and subject themselves to profound, prolonged, and sincere self-loathing. Only those able to atone for their wickedness could truly be deemed repentant and deserving of grace. They alone would enter the kingdom of heaven.

The same belief in man's weakness lay at the heart of Weyer's thoughts on witchcraft. The lonely old women who succumbed to the blandishments of the Devil, and the magicians who, spurred on by avarice and ambition, took the Devil's shilling, did not differ greatly from the rest of humanity. All were effortlessly tricked by Satan into committing the foul crimes to be expected of the weak and easily tempted humans of the Lutheran religious cosmology. Indeed, Weyer is likely to have drawn many of his beliefs about witches directly from Luther's writings. These were replete with discussions of Satan. Over and over again, Luther cited humankind's susceptibility to the Devil's wiles as a strong indication of its essentially base nature. He even recorded several personal encounters with Satan. Of one night-time debate with the Devil, Luther noted that he'd considered shouting at his adversary, 'I have also shit and pissed; wipe your mouth on that and take a hearty bite.' Luther was convinced that the Devil posed a constant threat.

More evidence of this is to be found at the end of Luther's autobiography. Believing that Judgement Day was at hand, he concluded with these stirring words, 'Reader, be commended to God, and pray for the increase of preaching against Satan. For he is powerful and wicked, today more dangerous than ever before because he knows that he has only a short time left to rage.' Believing implicitly in Lutheranism and fully committed to the cause of the Protestant Reformation, Weyer happily embraced Luther's insistence that the Devil is everywhere, becoming more active and desperate by the day.

Recognizing these three influences allows us to resolve the paradox of Johann Weyer. His unquestioning belief in the reality of black magic came first from Agrippa and then from Luther. And his humanity, probably a core aspect of his personality, was strongly reinforced by the medical humanitarianism of Agrippa and the all-weather compassion of Erasmus of Rotterdam. So it was that a man who saw the Devil as the author of virtually every misfortune and disaster could also defend witches as half-mad, half-possessed, but not even slightly to blame. Rather than being heroically ahead of its time, Weyer's outlook was entirely to be expected from a Lutheran of an Erasmian stripe writing just before the onset of the scientific revolution.

The myth of the mad witch

Weyer's reputation as the father of psychiatry hinges on another and much more pervasive misconception. It's one of the most popular tropes in the history of psychiatry that accused witches were often insane. They are said to have suffered from psychoses which led them to believe they had supernatural powers. Thus, reflecting on the elaborate confessions of the accused, many psychiatrists have retrospectively diagnosed hysteria or schizophrenia. Then they've represented the witch craze as a vicious pogrom against Europe's mentally ill by churches and states too hooked on superstitious dogma to perceive the power of scientific reasoning. 'Mental disorders,' writes one historian of psychiatry, 'because of the abnormal behaviour of sufferers, were definitely attributed to the maleficence of sorcerers and witches.' Weyer's heroic status rests on having reached the same conclusion long before anyone else. But this view of the witch craze, flattering as it is to the modern ego, doesn't stand up to scrutiny.

There is a certain naïveté in modern writers failing to see that there was nothing at all odd about people believing in powerful spirits, demonic possession, and magic spells during the 1500s. In a period in which few denied that God routinely intervened in the world, altering the course of nature for the betterment of mankind, this was hardly evidence of mental instability. The Church itself encouraged a belief in miracles, not least because these remained among the best arguments in favour of the truthfulness of holy scripture. Nor were the mysterious powers attributed to witches any weirder than the magical forces associated with the Catholic mass. If bread and wine were thought to be converted into the body and blood of Christ merely by a priest intoning a few special words at the altar, it didn't take much imagination to think that a witch could spoil crops or kill babies by invoking the name of Beelzebub. Witchcraft simply implied that the Devil could wreak as much havoc as God could do good. Both worked in mysterious ways.

More important, there's not even much evidence to show that those accused of witchcraft during the 1500s exhibited obvious symptoms of insanity. The profile of the suspected witch as a paranoid schizophrenic has tended to be based on accounts of witch trials in which crazed old women hurled spells from the dock, admitted the most improbable crimes, and maniacally predicted the ultimate victory of Satan. Yet historians typically drew examples such as these not from verbatim reportage but from the imaginative creations of hack writers. Striving to meet a fast-growing demand for titillation, sixteenth-century authors had to provide highly coloured accounts if they were to satisfy their readers' appetites for the sexually deviant, macabre, and picaresque. The frail and scared old woman who usually stood in the dock was of little interest. So, instead, the witch became a wild, sexually profligate beast, who stood in court fulminating wildly and, overcome with rage, cursed all and sundry in unknown tongues.

The most successful of these writers, like the editors of modern tabloid newspapers, quickly learned that a good story is one that teeters on the edge of believability. But behind the journalist's hyperbole lay the prosaic reality of frightened old women, cowering before the might of the law and the spite of their neighbours. Even the confessions extracted from the accused were of the most undramatic kind. The most common included 'killing swine', the sixteenth-century equivalent of the charge of 'economic sabotage' so beloved of those organizing Stalinist show trials. The death of the family pig could be a disaster for poor families, but it didn't imply the possession of elaborate powers.

Nor did these victimized old women seem to behave in ways suggesting psychosis or hysteria. Research now shows that the typical accused witch was little more than an old scold. Often single or widowed, she had a reputation for being quarrelsome, spiteful, and aggressive, and this was enough to marginalize her from her local community. Furthermore, the confessions of these women spoke not of mania but of a deep sense of social isolation. The Devil is said to have approached them when they were feeling sad, hungry, and terribly alone; Satan, they explained, promised them the solace they craved. Even though such confessions may have been obtained under torture, some have an air of authenticity. It may be that certain old women came to rationalize their sense of alienation by imagining that their souls had indeed been commandeered by Satan. Perhaps others claimed to have special powers in order to elicit a grudging respect from neighbours who would otherwise either bully them or ignore them entirely. All it then took was a seemingly inexplicable tragedy for accusing fingers to be directed towards whichever crone happened to be the most feared or despised.

Accusations of witchcraft reflected the normal tensions of village life in other ways too. Court records indicate that elderly women sometimes accused one another of being witches. In some cases the motivation may well have been material. Old women with neither husbands nor descendants to support them were reliant on handouts from their neighbours. The more poor widows there were, the thinner this charity had to be spread. So, in times of extreme need, it made economic sense for one or more supplicants to accuse others of witchcraft in the hope that their removal would mean more food for the survivors.

Stepmothers were also vulnerable. Children jealous of their father's affections sometimes claimed to have seen their stepmother praying to Satan, speaking to animal familiars, or sneaking off to partake in orgiastic sabbats. In lots of cases the accused were probably also suffering from depression, as Weyer claimed. But it was for their position within the community rather than their low spirits that they were victimized. In any case, when trial witnesses referred to the mental state of suspected witches they usually spoke only of sullenness and irritability. Given the lack of evidence to the contrary, it seems sensible to conclude that only a small minority experienced the grandiose delusions, hallucinations, or voices in the head that histories of psychiatry would lead us to expect.

Mythe d'origine

At first it's hard to imagine how the Weyer myth could have taken hold, given that most accused witches were neither hysterical nor schizophrenic, and that Weyer himself never doubted the existence of witches. Philippe Pinel, the pioneer French psychiatrist we'll meet in the next chapter, read Weyer's book in 1801. He immediately saw that its author believed in the reality of demonic possession. Taking a commendably balanced position, Pinel remarked that 'we should forgive' Weyer because he was subject to 'the common errors of his time'. It was only in the following decades that *De Praestigiis* came to be reread as a forward-looking classic. The key to understanding why this happened lies in the fact that nineteenth-century doctors were having an extremely difficult time trying to establish psychiatry as a medical specialty.

As we'll see in more detail in the next chapter, psychiatry began the century as an embryonic field with many rivals and only modest public support. As late as the 1830s, numerous European and American asylums remained in the hands of clergymen or, rather more commonly, lay caretakers. Even when doctors had finally managed to secure control of these, they continued to find it hard to convince the public that madness was the province of the medically trained. Worse still, despite the growing secularism of the age, many people persisted in believing in the possibility of satanic possession. Psychiatrists knew that so long as this kind of superstition survived, their profession would remain the poor relation of all other medical disciplines.

Just how much psychiatrists still had to do became painfully apparent to them with the unfolding of events in the small Alpine town of Morzine in June 1857. Morzine lies in a high valley of the southern Chablais, separated from Switzerland by a single mountain. In 1857 several of the town's adolescent girls claimed to have seen a vision of the Virgin Mary. The girls soon began to experience violent convulsions during which they 'relished blaspheming the Eucharist' and announced that they were the victims of a *mal donné*, a witch's curse. The local abbé tried to refute their fears, but eventually agreed to perform the rites of exorcism. This did not have the desired effect. Quite the opposite: by 1861 about 200 women and girls were claiming to have been bewitched or possessed by demons. News of the Morzine *mal* spread throughout France.

The region's senior clergy now stepped in and forbade further exorcisms, declaring the Morzine females to be deluded or attention-seeking. Denied the help of their Church, many of them walked to Switzerland in search of 'magnetizers' or priests willing to exorcise their demons. Observing from afar, French psychiatrists were appalled to discover that convulsions and hallucinations were still being blamed upon evil spirits. Perhaps seeing a chance to shine where the Church had failed, the psychiatrists now became involved.

First a physician from Lyons arrived in Morzine and minted a fresh diagnostic label for the occasion, hysterodemonopathy. But the townswomen steadfastly refused to accept that they were suffering from a medical condition, whatever it was called. And, as the numbers of women claiming to be possessed increased, a rather more influential psychiatrist was despatched.

Having journeyed all the way from Paris, Adolphe Constans immediately isolated or exiled the more vocal women and arranged for a small detachment of soldiers to be stationed in the town to impose order. It took a further three years for calm to be restored. Eventually, most of those who had been exiled or incarcerated were freed or allowed to return, and the soldiers were withdrawn. But then the bishop visited Morzine and about ninety women relapsed into swearing and screaming, and implored him to perform a collective exorcism. Order was restored only by recalling Dr Constans and the soldiers.

Over the following decade or so the Morzine affair gradually petered out. But the episode had brought home to the psychiatric profession just how superficial an impact their 'discoveries' had made beyond France's major towns and cities. Far too many people could still be persuaded that witches and demons actually existed and that only priestly exorcism could relieve the anguish of the afflicted. Determined to build on their comparative success in Morzine, the psychiatrists now stepped up their campaign to discredit supernatural explanations for hysteria and hallucination. As the historian, Patrick Vandermeersch, has shown, in this context Johann Weyer began his metamorphosis from being a little-remembered humanitarian to a heroic pioneer of modern psychiatry.

It began with a special lecture delivered at the Faculté de Médecine in Paris in 1865, during which one Theodor Axenfeld outlined Weyer's career. Shortly after, the eminent neurologist, Désiré Bourneville, produced a biographical sketch of Weyer, supposedly based on Axenfeld's talk. But whereas the latter had alluded to his subject's belief in the Devil, Bourneville refuted it, saying that Weyer was unquestionably *un impie*, an unbeliever. In fact, Bourneville's influential account expunged every mention of Weyer's belief in demons, and culminated in the following emotive claim:

> He strove to demonstrate that the crimes the witches were
> accused of were fictitious; that those women were not criminals
> but patients suffering from mental illness; that they should not
> be sentenced by priests, monks, nor judges; that they
> consequently should not be put in prison, tortured, and burned,
> but that they should be entrusted to the care of physicians.

The fabric of distortions, half-truths, and falsehoods woven by Bourneville has informed all but the most recent accounts of the history of psychiatry. But the Weyer myth survived so long largely because of its usefulness to the profession. Casting witches as harmless lunatics, and Weyer as an ultra-rationalist who risked his life by denouncing the superstitious hysteria of the Inquisition, provided psychiatry with an almost ideal *mythe d'origine*. By projecting their own values onto Weyer, psychiatrists were able to claim that their discipline was the culmination of a heroic conflict between reason and unreason.

The myth's real strength, however, came from the implicit association of science with enlightened compassion and, conversely, religion with ignorant barbarism. Set against a backdrop of punitive theology, it could be made to seem that the psychiatric approach to insanity was a triumph of both science and human decency. Using the Weyer legend, psychiatrists like Bourneville could then castigate the Church for murderous folly and, by implication, discredit any attempt by the religious of their own day to hold forth on matters concerning mental phenomena. Fired across the Church's bows, the story of Johann Weyer told clerics and their congregations that the Church had done enough harm already, and that next time they heard someone claiming to have seen a vision of the Virgin Mary, Jesus Christ, Lucifer, or St Peter, they had better call for the nearest psychiatrist. Bourneville himself had read *De Praestigiis* and must have known better. But he presumably felt that his carefully crafted distortions were serving the cause of a much bigger truth. The Church was wrong about madness, and society as a whole needed to be weaned off its misunderstandings.

Weyer's good deeds

Some would exonerate Bourneville on the grounds that tampering with the historical record helped to release mental illness from the grip of superstition. But a century on, that battle has been won, and it's now important that the record should be set right. Weyer was not a man centuries ahead of his time. His medical views were unexceptional and his belief in black magic profound. Nor, incidentally, does the witch craze give us material for a general critique of the role of the Church in history. Recent scholarship suggests that in most parts of Europe, established Churches acted as a brake on witchcraft accusations, and the worst pogroms occurred in those place where the authority of Church and State had broken down.

None of this is to say that Johann Weyer deserves to slip once more into obscurity. As an old man, he fondly hoped that posterity might pay him more respect than had the people of his own time. 'In a future age,' Weyer wrote, 'I hope men will rejoice that I lived.' Even if he cannot be considered the father of psychiatry, Weyer's condemnation of the 'savage execution' of witches warrants the fulfilment of his hope. Cicero and Tacitus are rightly respected for their atypical disgust at the horrors of the Roman games. So too should we regard Johann Weyer. To protest with the conviction he did against the witch-burnings of the fifteenth century required both courage and moral decency. It is for this strength of character that he should be remembered. To use a phrase coined by Weyer's near-contemporary, William Shakespeare, Weyer's writing of *De Praestigiis* deserves to be seen as a very 'good deed in a naughty world'.

PH. PINEL.

Prof.ʳ à la faculté de médecine de Paris. Médecin en chef de la Salpêtrière.
Né à S.ᵗ André, Dép.ᵗ de Tarn et Garonne le 20 Avril 1745.

The asylum looked like a circle of the Inferno when
Pinel entered upon his duties. The lunatics lay all
about, raving, riveted with chains and irons. They
were regarded as desperate, dangerous animals . . . He
planned to strike off the chains from these miserable
creatures and to inaugurate a regimen based on
kindness and sympathy.

    Albert Deutsch, *The Mentally Ill in America* (1967).

Behind a lattice of wooden scaffolding in June 1849 a large canvas was
hoisted into position. Measuring about 18 by 8 feet, Charles Müller's *Pinel
Orders the Removal of Iron Shackles from the Insane Men at Bicêtre Hospice* was
the second in a series of dramatic paintings of heroic medical deeds to be
hung in the lobby of the Paris Academy of Medicine. Müller depicted an
event still celebrated as the point at which the humane treatment of the
mentally ill supplanted the barbarism of earlier generations. The pioneer
psychiatrist, Philippe Pinel, is shown striding purposefully into the French
madhouse of Bicêtre in 1793, his immaculate white waistcoat resplendent
against the dull grey walls and the subdued figures of the insane. Terrified
inmates cower all around, some scarcely able to look in Pinel's direction,
others staring impassively at him, presumably resigned to another
'therapeutic' application of the whip, fist, or boot. Pinel, however, has
other ideas. Müller shows him directing one of his attendants towards a
weak and helpless old man, and orders his servant to unlock the heavy
iron chains by which the man is riveted to the wall. This instant, Müller
implies, marks the beginning of the end of centuries in which the insane
were abused, neglected, stripped of their humanity, and subjected to
whatever casual cruelties their custodians chose to inflict. Delivered into

LEFT: *Portrait of Philippe Pinel, a lithograph by Antoine Maurin.*

the humane hands of trained doctors, the mad were at last to have their dignity and perhaps even their sanity restored.

Müller's magnificent painting confirmed Pinel as one of the heroes of French medicine. 'One day Providence sent Pinel: his great genius brought about this transformation,' announced a Parisian newspaper the day after its unveiling. When, a few years later, Napoleon III created an Imperial Academy of Medicine, its entrance too was graced with an artistic homage to Pinel. A *rue Pinel*, a *place Pinel*, and a series of statues gave further testimony to the esteem in which the hero of Bicêtre was held. But the full story of the breaking of chains at Bicêtre has even more pathos than Müller's canvas conveys.

The official account tells how Pinel arrived at Bicêtre and was so disgusted at the treatment of the inmates that he marched off to remonstrate with the man in overall charge: the brutal, paraplegic creature of the new Jacobin government, Georges Couthon. Showing far more courage than prudence, Pinel stormed into Couthon's office unannounced and demanded that the insane be unshackled. We have to presume that he caught Couthon on a relatively good day. For, rather than having Pinel locked up, Couthon agreed to visit Bicêtre the following morning. At the head of a delegation from the Committee of Public Safety, he arrived early to see what he termed 'the wild beasts'. 'You must be insane yourself to want to unchain these animals!' he barked at Pinel before departing. But although Couthon's intention was clear, he hadn't expressly forbidden the unchaining. And so, risking his life by acting without Couthon's explicit authority, Pinel unfettered the asylum inmates. One of them, the tall and imposing M. Chevingé, showed Pinel his gratitude by remaining his faithful servant to the end of the good doctor's life.

Pinel, revolutionary hero

This story has made Pinel much more than just a hero of medicine. He has come to stand for all that was noble about the French Revolution before its lofty idealism was suborned first by the Terror and then by Napoleon's ruinous bellicosity. It's for this reason that in 1885 Bastille Day was chosen for the unveiling of the most famous statue of Philippe Pinel. Created by Ludovic Durand, it stands at the entrance to the women's asylum, Salpêtrière, where Pinel served after leaving Bicêtre. Paid for by public subscription, Durand's masterpiece shows Pinel holding the links of a broken chain in his hands as two young girls gaze up at their liberator in mute gratitude. A day sacred to the memory of the Revolution, 14 July, was deemed perfect for the commemoration of one who typified its early promise.

The chains held in Pinel's hands were themselves invested with immense symbolic significance. Chains were after all one of the principal motifs of the French Revolution. Jean-Jacques Rousseau's 1762 book, *The Social Contract*, opened with the immortal lines, 'Man is born free, and everywhere he is in chains.' And Pinel's act was made an even more potent symbol because the shackling of lunatics had come to stand for the dehumanizing bondage endured by the millions of have-nots who laboured under the Bourbon monarchy. The breaking of the Bicêtre chains was the Revolution in microcosm.

The event was also symbolic of the fight for *liberté, egalité,* and *fraternité* in a more specific sense. Prior to the French Revolution it was widely believed that many of those incarcerated in the nation's asylums had been only too sane on entry. These were the victims of the *lettres de cachet*, infamous documents by which the King of France imposed his arbitrary will on his subjects. Made effective by his small seal (*cachet*), the *lettres* secured the incarceration of political enemies, denying them both trials and formal means of appeal. Not only the king, however, had access to the *cachet*. It was well known that, for a price, palace officials would make these *lettres* available to those who wished to obtain more than their fair share of a family inheritance by having other beneficiaries locked up in prisons or asylums. In the years leading up to 1789, the *lettres de cachet* came to epitomize the vicious and extravagant corruption of the old regime. By extension, Pinel's removal of the fetters in Bicêtre was a hammer-blow against Bourbon oppression.

What gave Pinel's story added punch in 1885 was that he had himself started out as a passionate republican. Having qualified as a doctor at Toulouse, he came to Paris in 1778 and rapidly made friends with a radical set of physicians, including Pierre Jean Georges Cabanis

Pinel fait enlever les feus aux aliénés de Bicêtre ('Pinel orders the removal of iron shackles from the insane men at Bicêtre Hospice') by Charles Müller.

and Antoine Fourcroy. Both men would play central roles in reforming medicine after the Revolution. But in the 1780s Pinel's association with them destroyed his chances of gaining a medical scholarship from the panel of royalist physicians who, complete with powdered wigs, gold-topped canes, and signet rings, decided his future. As a result, Pinel had to support himself through hack journalism and teaching mathematics.

Only after the Revolution did Pinel's fortunes change. Cabanis and Fourcroy made sure that his early support for the republican cause was well rewarded; hence his appointment first to Bicêtre and then to the Salpêtrière. It is an indication of Pinel's profound humanity that his radicalism did not survive the guillotining of Louis XVI. None the less, his reputation as a die-hard republican stuck fast. So, when the Bourbons returned to power after the Battle of Waterloo, Pinel was one of many leading physicians to fall out of favour. The tide continued to move against him, and on 2 February 1823 he was forced from his post and reduced to half-pay. Three years later he died. Want of courage, political antagonism, and indifference meant that his passing went largely unmarked by the medical profession. But in time this rebounded to Pinel's posthumous advantage. His shoddy treatment under the restored Bourbons allowed latter-day republicans to celebrate him as one of their own, his apostasy quietly forgotten.

Within half a century his position had been transformed. Since the 1880s Pinel has been considered the father of French psychiatry and the removal of the chains in Bicêtre the moment at which psychiatry proper was born. However, as we have come to expect in such cases, it is as well to remember Oscar Wilde's famous aphorism: 'Truth is rarely pure and never simple.' For in recent decades, historians have shown that the event celebrated by Charles Müller's painting never actually took place. In every detail, the account of Pinel having the chains removed in Bicêtre is pure invention. The inmates of Bicêtre *were* eventually unshackled, but not by Pinel and emphatically not on his orders.

Yet the spuriousness of *le geste de Pinel* is neither the most curious nor the most revealing aspect of this story. Far more remarkable is the fact that while Pinel *was* a genuine pioneer of humane psychiatry, nearly everything he achieved for the benefit of patients was quietly suppressed by the generation of psychiatrists that followed him. The real Pinel, who insisted that patients ought to be consoled and encouraged rather than bullied and bled, was buried beneath a myth by men who found his ideas incompatible with their personal and professional aspirations. Paradoxically, far from being elevated by the myth-makers, Pinel was in fact diminished. Why he was remembered only for something he didn't do is the subject of this chapter. Our first task, however, is to chisel away the myth.

Pussin and the Bicêtre chains

In 1772, Jean-Baptiste Pussin's career as a tanner came to a premature end. He had contracted scrofula, a form of tuberculosis causing large and painful swellings in the region of the neck. Pussin sought medical help at Bicêtre, which was what we would now term a general hospital. Having made a full recovery, he then applied for a job on the hospital staff. Having been appointed, Pussin proved a talented worker. So, after a short while, the hospital administrators placed him in charge of the St Prix ward for the mentally ill. Monsieur Pussin, and the equally able Madame Pussin, quickly acquired reputations for tolerance and efficiency.

The conventional image of the eighteenth-century asylum is of a sink of despair and degradation in which the insane were treated like criminals or the wild beasts to which Couthon is said to have referred. This picture has some validity. When a committee of British Members of Parliament enquired into the conditions at London's Bethlem Hospital in the 1790s, they heard shameful accounts of filthy wards, madmen who spent their lives in chains, a physician who was never there, and a resident surgeon who for ten years had been 'generally insane and mostly drunk ... so insane [in fact] as to have a strait-waistcoat'. And, despite the efforts of the inspectors, another official visit to Bethlem in April 1814 revealed dozens of patients cramped, chained, virtually naked, and lying on soiled straw. One of them, an American seaman, was confined to a cell and had been pinioned to a wall for twelve years, his freedom of movement limited to a twelve-inch arc. The appalled inspectors reported that the insane were being treated like 'vermin'. Nor was Bethlem an isolated exception. As late as 1825, there was a healthy trade in books with lurid titles such as *The Crimes and Horrors in the Interior of Warburton's Private Madhouses at Hoxton and Bethnal Green.*

London's Bethlem and Warburton's establishments were not, however, representative of all asylums. And they were certainly not reliable indicators of the conditions within Pussin's St Prix ward. When the duc de La Rochefoucauld-Liancourt conducted an official inspection of the Bicêtre asylum in May 1790, he was able to report that 'only ten out of 270 [patients] were chained,' and these were the violent minority. In short, the situation at Bicêtre was nowhere near as horrific as Charles Müller's painting portrays it.

Pussin's and Pinel's paths crossed only after the Revolution. Pinel had already gained some experience with the mentally ill by working at a private sanatorium for the insane owned by a former cabinet-maker. Then, on 6 August 1793, thanks to the good offices of Cabanis and Fourcroy, he arrived at the Bicêtre asylum in the capacity of 'physician of the infirmaries'. As a result, Pussin became one of his subordinates. According to the traditional account, Pinel had hardly had time to remove his overcoat before marching off to remonstrate with the wicked Georges Couthon. But we now know this to be a complete fabrication. Chains continued to be used to restrain a dozen or so prisoners throughout the two years Pinel worked at Bicêtre.

It was only in May 1797, two years after Pinel's departure, that Pussin did what could have been done any time in the preceding years: abandon the chains and use the much more humane straitjacket to restrain violent or suicidal patients. Doing so had apparently never occurred to Pinel himself. And the story of his brave appeal to Couthon is no more than an elaborate invention. Recent research has revealed that Couthon was out of town in the weeks when his famous visit to the 'wild beasts' of Bicêtre is supposed to have taken place. Perhaps more surprisingly, when Pinel heard of Pussin's decision to do away with chains and iron staves, he was markedly slow to follow suit. Only three years later, with Pussin now his loyal assistant at the Salpêtrière women's hospital for the insane, did Pinel opt for a more humane means of restraint. With the assistance, and no doubt the encouragement of Monsieur and Madame Pussin, he finally sent for the keys.

But even this didn't lead to a general improvement in the care of the insane. Pinel's disciple, Jean-Etienne Esquirol, undertook a nationwide survey of asylums in 1819. His testimony vividly demonstrates that the work begun at Bicêtre and the Salpêtrière in the 1790s remained tragically incomplete. Twenty years later France still had dozens of asylums as bad as Bethlem and, in some cases, worse. As Esquirol wrote:

> I have seen them [i.e. the insane] naked, clad in rags, having but
> straw to shield then from the cold humidity of the pavement
> where they lie . . . I have seen them at the mercy of veritable
> jailers, victims of their brutal supervision . . . in narrow, dirty,
> infested dungeons without air or light, chained in caverns
> where one would fear to lock up the wild beasts that luxury-
> loving governments keep at great expense in their capitals.

Playing advantage

So how did a supporting character in a very modest victory for the insane come to be reinvented as the central figure in an epic drama? It all began with an impecunious son. In 1823, having lost his job, Pinel suffered a series of incapacitating strokes. He was not the only one to suffer from the resulting drop in income. His eldest child, Dr Scipion Pinel, himself a psychiatrist, was heavily in debt and had already used his father's property and reputation to obtain more cash. As a stop-gap he next plundered his father's library, but what he really needed was some means of permanently restoring his fortune. It was under these circumstances that he came up with the idea of recasting his father as one of the major heroes of psychiatric medicine. Not only would the son of the hero of Bicêtre reap rich professional rewards by practising in the same field, he would also be in a far better position to borrow money.

With his father unable to protest, in 1823 Scipion published a short article on 'The Removal of the Chains', which he falsely claimed to have found among his father's papers. Describing how Philippe had ordered the abandonment of chains at Bicêtre, he made no mention of Pussin. Over the following years Scipion Pinel continually embellished the story until, in 1836, he gave the now generally accepted account during a presentation to France's Royal Academy of Medicine. With all the main protagonists now dead, he had a much freer rein to invent. His master-stroke was to write in the wicked republican, Georges Couthon. Scipion judged the political situation well. In 1830 Louis-Philippe d'Orléans had become King of the French. Although he was no reactionary, Louis-Philippe had an understandable distaste for all things republican. The new scientific elite, largely made up of ardent royalists and men who took their religion seriously, shared this antipathy. One of the most powerful was Georges Cuvier, a brilliant comparative anatomist who claimed to be able to reconstruct an entire fossil organism from a few teeth and fragments of mandible, and who had mercilessly destroyed the career of the early evolutionist and deist, Jean-Baptiste Lamarck. So, in characterizing Couthon as vicious and corrupt, Scipion ensured for his story a favourable reception among royalist cliques. Pitting his father against a notorious Jacobin like Couthon also helped to sever the associations between Pinel and the Revolution. There is a certain irony here. Philippe had proudly dubbed the infant Scipion his 'little republican'. It was not a prophetic choice of words.

Then Jean-Etienne Esquirol, Pinel's one-time student and disciple, put his shoulder to the wheel by including a version of the chain-breaking myth in his enormously influential treatise on mental illness. Esquirol was striving hard to have conditions in asylums improved, and his campaign needed a heavyweight hero. But he too saw the need for a little spin. Hushing up Pinel's republicanism, Esquirol stated, 'The ideas of the time perverted the importance of breaking the chains that degraded and troubled the madmen at Bicêtre.' Science and politics were inseparable in nineteenth-century France, and Esquirol was astute enough to realize that if the royalists were to accept the Pinel myth, a bit of fudging was necessary. The strategy worked, for it was during Louis-Philippe's reign that Müller's painting was commissioned. Politically sanitized, Philippe Pinel once more became a hero of French psychiatry.

But what is most striking about the myth constructed by Scipion and propagated by Esquirol is what it left unsaid. As we will now see, on the basis of what Pinel actually did at Bicêtre and the Salpêtrière, one could really make a case for his being a major player in the development of

French psychiatry. The machinations of his son, and others, meant that both he and Pussin were deprived of the respect they were due.

'The ways of gentleness'

Although Pinel was strangely unconcerned about the continued use of chains at Bicêtre, his general approach to psychiatry was unquestionably humanitarian. He vehemently rejected the widespread belief that routine beatings help to cure insanity. 'Maltreatment,' he wrote, 'or the ways of a too harsh repression, exacerbates the illness and can render it incurable.' He also had little time for the battery of available medical 'treatments': Pinel dismissively referred to bleeding, emetics, purgatives, electrical shocks, freezing cold baths, and rotating chairs as 'polypharmacy'. He also condemned the keeping of the mad in insanitary conditions and tried to give Salpêtrière the atmosphere of a comfortable home, complete with uplifting books, drawing implements, musical instruments, and quiet places for walking to distract patients from their troubling and obsessive thoughts.

Above all, Pinel dedicated his career to introducing a therapeutic approach known as 'moral treatment', which rested on 'the ways of gentleness'. 'Lunatics,' he wrote, 'should not be regarded as criminals, but as diseased persons inviting our utmost compassion, and whom we should seek by the most simple means to restore to reason.' For Pinel the effective treatment of insanity required the doctor to form an intense, supportive, and respectful relationship with the patient, through which the latter's reason and self-control could be restored. The insane were not, he insisted, 'absolutely deprived of reason' but could be helped to overcome their difficulties by a doctor who had so gained their confidence through 'consoling words' that he could 'change the vicious chain of their ideas'.

In one of the cases Pinel recorded, a young man who'd had a couple of bouts of insanity appealed for help, 'tormented by the fear of an imminent relapse'. Pinel placed him in the charge of the young Esquirol, who immediately began developing a rapport with the patient. Soon after, a new crisis came on and the patient became convinced that he was about to plunge once more into madness. Employing 'consoling words' and 'forceful language in order to rouse his battered courage', Esquirol hardly left the patient's side till the danger had passed. Shortly after, a much improved patient congratulated Esquirol with the words, 'You have saved me,' and spent the evening contentedly playing billiards.

Pinel was too much of a realist, though, to believe that gentle words could solve everything. Some patients were simply too hostile or unresponsive to take heed. He had two solutions in such cases. The first was borrowed from the Reverend Dr Francis Willis, who ran a private asylum in Lincolnshire and was famous for having treated George III. His 'technique' was to overawe patients by turning upon them an extraordinarily 'piercing gaze' and then gradually to bend them to his will. During a Parliamentary enquiry into his treatment of George III, Edmund Burke asked Willis if he really thought it safe to give the king a sharp razor during his morning ablutions. What would Willis have done had the king turned violent? Willis unhesitatingly responded that he would have pacified him 'by the EYE! I would have looked at him *thus*.' Burke immediately looked away, humbled by the power of the reverend gentleman's gaze. Pinel saw merit in this approach. And, despite being a somewhat shy man, he too argued that it was sometimes 'necessary to subjugate first, and encourage afterwards'.

Pinel's other method was more theatrical and designed to so 'shake up' the patient's 'imagination' that their reason was jolted back into life. In 1794 a tailor arrived at the Bicêtre hospital, convinced that he was under sentence of death by guillotine, a delusion that appeared soon after Louis XVI had received the same sentence. The tailor had made the mistake of voicing his reservations about the latter judgement in the presence of fierce republicans in his quarter. Thereafter he was treated with suspicion and, several days later, overheard some menacing words that he believed concerned himself. The tailor returned home in a fearful state, shaking, almost prostrate with paranoia, unable to eat or drink, and utterly convinced that he was to be executed as an enemy of the Revolution.

After several months under Pinel's care, the man's delusion began to subside. Then he relapsed, so Pinel tried a more radical course of therapy. Three young doctors were dressed up in the black robes of magistrates and arrived at Bicêtre, claiming to have been sent by the *Corps législatif* to investigate the tailor's political views. Having thoroughly interrogated him, the verdict was pronounced: acquittal. 'We acknowledge,' the pretend magistrates stated, 'having found in him only the sentiments of the purest patriotism.' The tailor's symptoms abated, at least temporarily. Whether this should be seen as straightforward trickery rather than psychiatric medicine is a moot point. Moreover, at this stage of the Revolution the tailor's fears might have been well grounded. But Pinel's desire to do whatever possible to help the patient recover could hardly be more evident.

Like modern psychotherapists, Pinel recognized the importance of building up a close, collaborative relationship with patients. By talking and consoling he sought to lift the spirits of melancholics and to correct the delusional state of mania. He also insisted that the insane, whether rich or poor, be accorded the dignity of medical patients. While what he termed his 'moral treatment' was hardly new, he championed the rights of the insane with a zeal and humanity unrivalled among his professional peers. And in terms of changing attitudes towards madness, few books in the history of psychiatry have been as influential as Pinel's 1801 *Traité médico-philosophique sur l'aliénation mentale ou la manie*. So why did posterity choose to forget this solid achievement and instead 'remember' an event that never took place?

Sins of the fathers

When Scipion Pinel began beatifying his father in 1823 he said nothing of the lifetime of effort Philippe Pinel had invested in demonstrating the value of moral treatment. Ten years later Esquirol was no less reticent. He mentioned Pinel only in relation to his fictional unchaining of the Bicêtre insane. Esquirol's motivation may, in part, have been personal. Leaving Pinel out of the picture enabled Esquirol to garner much of the kudos for having bestowed dignity on the insane. But there was also a much wider professional interest at stake. By the 1820s, Pinel's exclusive emphasis on the 'ways of kindness' had become acutely embarrassing to a psychiatric profession trying to assert its primacy in the area of caring for the mentally ill but finding its rivals exasperatingly hard to evict.

Philippe Pinel had been quite candid in admitting that he was heir to a tradition initiated not by respectable medical men but the charlatan, the cleric, and the concierge. As this implies, Pinel's ardent republicanism had chimed with his approach to psychiatry. For although he believed that psychiatrists could lay claim to specialist knowledge, he insisted that physicians had more to learn from the quack and the lay custodian than from the stuffy, robed clique of the old order. In making this point Pinel indulged in plenty of crowd-pleasing invective. Praising the medically untrained, he wrote in 1801, 'Men who are strangers to the principles of medicine, guided only by sound judgement or some obscure tradition, have devoted themselves to the treatment of the insane, and they have effected a great many cures.' Many of these untutored experts were the concierges who, like Pussin, had been placed in day-to-day charge of the asylums.

Pinel believed that by living 'constantly in the midst of lunatics' such people had acquired far greater practical expertise in the treatment of madness than had the medical elites of pre-revolutionary days. 'Full of doctoral pomposity', he inveighed, the elite physician had made 'transitory visits' to asylums and left with an 'exclusive confidence in his own knowledge'. But of what value was such knowledge in comparison to that of concierges who were 'continually' exposed to the 'spectacle of all the phenomena of insanity'?

Leading by example, Pinel freely admitted his debts to the plebeian Pussin. Pinel also heaped praise on the Tuke family of York. In 1796, horrified at the appalling treatment of the insane under the auspices of the physicians in the York asylum, the Quaker tea-merchant, William Tuke, had opened up his own asylum called the Yorkshire Retreat. Entirely run by sympathetic laymen who, among other things, rejected the use of chains, it sought to create a family atmosphere for patients. Tellingly, the Retreat did not have a resident physician.

Pinel's admiration for Francis Willis also illustrated his non-medical loyalties. Willis was an ordained clergyman who had only taken a medical degree after being bullied by the medical authorities into doing so. By then, however, he had already spent several years treating the insane and had found the prevailing medical strategies worse than useless. What he did find of great value was his own training and experience as a vicar.

Although Pinel made nothing of it in print, this was a background he too shared. For several years, until his twenty-fifth birthday, Pinel had studied at the seminary of the Fathers of the Christian Doctrine in Lavaur and was expected to take holy orders. It is therefore no surprise that Pinel's later moral treatment has been described as sacerdotal psychiatry. His emphasis on exercising a gentle authority over each of his patients echoed both the priest's role in the confessional and the attentive manner in which novices were trained by the Fathers of the Christian Doctrine. The same Fathers taught the value of caring for the poor and the dispossessed. Pinel's decision to minister to the insane, his devotion to poor patients, and his practice of calming patients in their final hours all testify to the continued influence of his religious education.

Pinel's claim that madness rested on a perverted understanding that could sometimes be corrected with the help of theatrical ploys drew on important philosophical ideas. But it too had much less esoteric associations. In the wake of the Revolution, a new form of civic celebration called the *fête révolutionnaire* was introduced into France. Blatantly propagandist in intent, the organizers of these fêtes believed that by bombarding viewers with certain impressions, they could imprint upon their minds a connection between the Revolution and the prospect of future happiness and riches. It was an attempt literally to mesmerize the public into an unwavering support for the revolutionary regime.

Pinel's psychiatry, then, had its roots in Catholicism, charlatanism, and mesmerism. He had attempted to put the field on a more scientific basis by producing classifications of madness and publishing state-of-the-art statistics on its causes. But to his son's generation, Pinel senior's version of psychiatry was far more redolent of pastoral care than of a rigorous medico-scientific discipline. Men such as Scipion Pinel and Esquirol never denied that being nice to patients was an essential part of proper treatment. Yet for them this wasn't enough; not, it must be said, because they believed it short-changed the patient. Top of their agenda was something rather different. Both men were at the forefront of a campaign to have France's public and private 'madhouses' brought exclusively within the compass of medical authority. Philippe Pinel's religious antecedents and his fondness for the dedicated layman were of no help to them at all.

For centuries asylums had been seen as custodial repositories for the mad, not centres of curative excellence. For society it was a case of 'out of sight, out of mind' or, perhaps more appositely, 'out of mind, out of sight.' The most patients could hope for was effective administration. The emergence of moral treatment in the 1790s at last made it seem that asylum regimes could help to restore a patient's sanity. But, unfortunately for physicians keen to enlarge the medical sphere, being kind and consoling required no specialized medical training at all. If gifted with the right temperament, the clerics and lay personnel who had traditionally run asylums could do this at least as well as physicians. And here was the rub. Scipion and Esquirol immediately saw that Pinel's commitment to 'the ways of kindness' had made it even harder for psychiatrists to squeeze their rivals out. To the deep discomfort of fellow professionals, Pinel's published works had lavished praise on concierges and clerics. In so doing, he'd given medical legitimacy to the rights of tenure these people had already acquired in French asylums.

A three-horse race

In the late 1790s, those who had pushed the idea of the moral treatment had made no bones about the limited role they saw for the medical profession. A Monsieur Broutet, for instance, explained that at his hospice for the insane in Avignon, physicians were only called in to prescribe 'physical remedies'. Broutet proposed setting up lots more of these lay-administered hospices. 'The administrators,' he added, 'to whom the lunatics would be entrusted would not have to take the trouble of becoming health officers as far as moral remedies are concerned.'

A similar proposal was made by the former cleric François Simonnet de Coulmiers, director of Charenton, one of the flagship Parisian asylums. Coulmiers left the resident physician in no doubt as to who was in charge, refusing to let him anywhere near patients unless they had fever, vomiting, or diarrhoea. This so alarmed the medical establishment that on the death of the physician who served Charenton, they foisted on Coulmiers a medic, Antoine-Athanase Royer-Collard, who (not least because he was the protégé of a divisional chief of the Ministry of the

Interior) submitted far less willingly to the ex-cleric's authority. Royer-Collard verbally bludgeoned Coulmiers into submission. Making abundantly clear where the real authority for looking after patients lay, he wrote to Coulmiers. 'It is to me and not to you that the patients are entrusted ... As for you, you are charged with furnishing [them] with food, bedding, laundry, and blankets, and keeping them clean.' In desperate response, Coulmiers snidely wrote to the minister of the interior claiming that Royer-Collard was a dangerous traitor to the Bonapartist cause and ought to be arrested. Unsurprisingly, given Royer-Collard's connections, Coulmiers' letter was ignored. Similarly unpleasant little duels were being fought out in asylums throughout France and England; many lay concierges fought a prolonged rearguard action against the steady encroachment of the medical fraternity.

For psychiatrists, clergymen were just as much a concern as the concierges, since many asylums also had resident priests. 'What ought to be the relation of the priest to the physician and to the lunatics?' asked one psychiatrist, rhetorically, as late as 1845. 'Is it not to be feared that the priest will usurp, even destroy the authority of the physician?' By way of analogy, this was like an American settler speaking of Native Americans as trying to 'usurp' ownership of the land on which he had settled. Human nature is such that, all too soon, interested parties forget who the incumbent is and who the incomer. But there can be no doubt that this collective amnesia was advantageous to the psychiatrists. For, like the white settlers of the American plains, they had a real fight on their hands. As the historian Jan Goldstein puts it, how could one have 'two father figures, two bringers of consolation, two healers of the spirit who employed moral means' in an asylum and not end up with a bitter contest for authority? 'Will the priest submit to medical authority?' one worried doctor was led to wonder.

To make matters worse, psychiatrists were also having a hard time persuading the general public that madness was a medical condition that had nothing to do with evil spirits or divine retribution. The problem with Pinel's approach was that it emphasized intangibles: thoughts, impressions, feelings, and delusions. It was too easy for such phenomena to be understood in theological terms as corruptions of the soul or the immaterial mind. This was an especially acute problem for French psychiatrists because the Catholic Church, after a brief eclipse during the Revolution, had regained immense social and political power. The Bourbons, Louis-Philippe, and Emperor Napoleon III all sought to restore to Roman Catholicism much of the ground it had lost during the Revolution. Challenging the authority of the Church was not something to be undertaken lightly.

As the physicians prepared for battle, it became ever clearer to them that Pinel and his writings were part of the problem. A man who questioned medical insights, favoured clerics and lay personnel, and concerned himself with immaterial imponderables was no friend to those who, in Esquirol's words, were working to invest the psychiatrist with 'an authority that no one can subtract himself from'. For all Pinel's dedication to his patients, the fact was that he had spent twenty years making Esquirol's *coup* harder to pull off. So his career had to be recast. Away went his entire life's work and in its place was erected a totally spurious myth. This masterpiece of reconstruction completed, Esquirol and Scipion then set about redefining madness and its treatment in a way deliberately inimical to the interests of the layman and the cleric.

Tackling things head-on

While never abandoning the moral treatment, Esquirol and his colleagues embellished it with a collection of theories and techniques serving to make madness a malady and its treatment a medical art. At the forefront was the new discipline of phrenology, which first appeared in the 1790s. It seemed to offer the possibility of once and for all shifting the perceived seat of madness away from the immaterial mind, the preserve of clerics, to a material brain, the province of the physician. Franz Joseph Gall, a Viennese physician, argued that all the different mental functions are localized in particular sections of the human brain. Everything from intelligence to valour, the impulse to procreate or the impulse to murder, the gifts of metaphysical perspicuity or of wit could, Gall said, be traced to discrete areas of cerebral grey matter. What's more, the size of these compartments, as reflected in the conformations of the skull, was believed to provide a faultless guide to the individual's talents and character.

'Feeling the patient's bumps' became shorthand for a new medical specialism within an emergent science of the mind.

Although later associated with itinerant quacks, phrenology first met with professional acclaim. George Combe, Britain's leading exponent of Gall's doctrines, didn't exaggerate the explanatory power that phrenology was widely seen to have: 'it is peculiarly fitted to throw a powerful light [on] Education, Genius, the Philosophy of Criticism, Criminal Legislation, and Insanity.' Accordingly, many leading French, British, and American psychiatrists, not least Esquirol himself, embraced aspects of Gall's theory. As the historian of medicine, Erwin Ackerknecht, put it, '[phrenology] was at least as influential in the first half of the nineteenth century as psychoanalysis in the first half of the twentieth.'

The metamorphosis of Philippe Pinel helps us to understand why so many highly gifted men adopted a set of ideas that would later be abandoned to mountebanks and fairground sideshows. In the absence of something better, any theory that said madness was due to physical deficiencies in the brain would find favour among psychiatrists. For once the brain was seen to be just another organ—one that secreted thoughts much as the gall bladder secretes bile—madness and the institutions for its treatment would become the exclusive domain of the psychiatrist. The cleric could be chased back to his altar and the lay custodian demoted to the role of a functionary.

Pinel had not been immune to the intense excitement aroused by phrenology. Purchasing a pair of callipers, he spent several months carefully measuring the skulls of his patients and comparing the results with the supposedly perfect cerebral dimensions of the statue of Apollo in the Paris museum. But here too, he proved a disappointment to his followers. Less than impressed by his findings, Pinel continued to emphasize the importance of purely 'psychological' causes: failed love affairs, financial ruin, religious fanaticism, and over-ambition.

In time, Pinel's scepticism about phrenology became more widespread. Many doctors concluded that skull shape would never be a useful guide to insanity. After all, madness usually struck years after the skull had hardened, so that the 'before' and 'after' skulls of the insane were identical. In short, behaviour could not be perfectly correlated with skull topography. But this didn't lead to the abandonment of attempts to 'medicalize' insanity. Instead, the gradual eclipse of phrenology inspired physicians to carry out thorough investigations of the brain itself. Rather than fumbling on the surface, the next wave of investigators opened up the crania of patients who had died insane and looked for characteristic changes in the brain's spongy convolutions. This approach was synchronized with wider trends in French medicine. Following the Revolution, a law had been passed stipulating that any patients who died in hospital unable to pay their bills had to leave their cadavers to medical science. No longer reliant on a trickle of bodies supplied by grave-robbers or bought from executioners, physicians could now begin to look deeper into the anatomy of disease. Doing so showed that external symptoms, such as fevers, rashes, swellings, or oddly coloured stools, were often correlated with damage to particular internal organs. Eager to see if the external symptoms of madness were also linked to physical lesions in their own organ of interest, psychiatrists extracted, sectioned, and studied their dead inmates' brains. This was the origin of a new pathological anatomy of the brain.

Bayle's brains

In 1822 Antoine Laurent Jessé Bayle, a nephew of one of Pinel's students, published the results of almost 200 autopsies conducted under the watchful eye of Athanase Royer-Collard at the Charenton Hospice. Six cases were particularly striking. In each one the patient had died after suffering from severe paralysis and dementia. On disinterring their brains, Bayle found that a membrane that stretched around them was severely inflamed. This inflammation, he concluded, was the 'cause of a symptomatic mental alienation'. Insanity, it seemed, was just another medical condition: much as inflammation of the lung membranes caused pleurisy, swollen brain membranes produced the psychological symptoms of madness.

Some of Bayle's colleagues correctly pointed out that, while he had established a correlation, he had not proved causation. Even supporters such as Esquirol felt that he was overreaching himself in claiming that virtually all cases of madness derived from some kind of swelling in the brain. On one thing, however, both critics and supporters were agreed. As Etienne Jean Gorget, the most caustic and determined of the former put it: 'The treatment of madness must thus be founded on the state of the brain.' The British psychiatrist, W. A. F. Browne, took a similar line. In *all* cases of madness, he asserted, proper post-mortem examination would reveal 'in or around the brain . . . some obvious alteration of structure.' The only difficulty here is that this simply isn't true.

In fact, at some level Bayle, Gorget, and Browne must have been aware that the vast majority of autopsies failed to find anything physically unusual about the brain at all. Once again Pinel was on the right side of the argument and on the wrong side of the politics. Observing that 'derangement of the understanding is generally considered as an effect of an organic lesion of the brain,' he pointed out that 'in a great number of instances, [this is] contrary to anatomical fact.' A British writer, William Nisbet, made the same point: 'In three fourths of the cases of insanity,' he noted, 'where they have been subjected to dissection after death, the knife of the anatomist has not been able, with the most scrutinising care, to trace any organic change to which the cause of disease could be traced.' Indeed, of Bayle's 200 eviscerated brains, all but six were anatomically uninteresting. Self-evidently, this was no basis on which to erect a bold new medical paradigm; but, self-servingly, this is exactly what the embattled psychiatrists did. And this post-Pinel generation, determined to reduce the cause of insanity to damaged cerebral tissue, was happy to labour under a collective misapprehension.

Of leeches and rotating chairs

Thinking of madness as the result of lesions in the brain had another advantage for psychiatrists. It allowed them to claim that the mind could only be treated by working on the body: reducing inflammation, clearing blockages, improving circulation, and the like. Pinel had been unimpressed by the curative powers of traditional remedies. 'How many patients do we see cured,' he wrote in 1801, 'without having been bled, and how many have been bled and remained incurable!' Here Pinel had been marching to the medical Marseillaise of the revolutionary era. While

British physicians continued to dispense cure-alls that caused their patients to salivate uncontrollably or lock themselves in the privy for hours at a time, Parisian physicians typically argued for non-intervention. 'The British kill their patients; the French let them die' was a popular saying. But even in Britain in the early 1800s, confidence in medical treatments of insanity was at a low ebb. When confronted by a government commission in 1816, one of the medical staff of Bethlem Hospital admitted, 'this disease is not cured by medicine,' and added reluctantly, 'If I am obliged to make that public, I must do so.'

This posed a very serious problem for physicians on both sides of the Channel. Admitting that their drugs and other remedies didn't work made it very hard for them to claim that madness ought to be within the jurisdiction of the medically trained. In Britain, Samuel Tuke defended his running of a madhouse without the help of physicians on the basis that 'pharmaceutical means have failed.' Another Quaker philanthropist, Edward Wakefield, horrified psychiatrists by commenting that they were 'the most unfit of any class of persons' to treat the insane because 'medicine has little or no effect on the disease.'

By seizing on the scant autopsy evidence furnished by Bayle and others, psychiatrists of the 1830s declared themselves a lot more sanguine about the usefulness of medicine. As it had been 'proved' that insanity was organic, it seemed that medicine simply had to work. If brain inflammation caused madness, then anything known to reduce swellings was bound to do the trick. Thus, the leading British surgeon and psychiatrist, J. G. Millingen, saw reducing cerebral inflammation as among his most important duties. His preferred method was 'moderate blood-letting' which involved attaching several small leeches to the 'temples and back of neck'. Millingen also recommended a varied regime of hot and cold baths, intended to promote good circulation, and a diet designed to 'keep the bowels regularly open'.

The American psychiatrist Benjamin Rush nearly always favoured venesection. 'The first remedy,' Rush insisted, 'should be blood-letting. It should be copious on the first attack ... From 20 to 40 ounces of blood may be taken at once.' Rush found 'the effects of this early and copious bleeding [to be] wonderful in calming mad people'. (This is hardly surprising; having lost 40 ounces of blood even the most athletic would feel all too literally drained.) Other 'remedies' in regular use by psychiatrists were blistering, purging, salivation induced by mercury, opiates, cold showers, electric shocks, gyration and swinging (using various mechanical contraptions, digitalis, prussic acid, camphor, rectified oil of turpentine, and that 'most universally agreeable substance', tobacco. Each psychiatrist had his own favourite therapies, which he extolled with absolute confidence. But well into the second half of the century blood-letting remained the most popular. Faced with this battery of treatments, no doubt many patients would have given much to have been assigned solely to the care of a kindly cleric or a latter-day Pussin.

Of course, now we know that most of these remedies were, at best, useless. In terms of medicalizing mental health, however, the effectiveness of any given treatment was not the most important thing. The key to success was confidence on the part of psychiatrists that their remedies did work. And convincing themselves that the treatments *were* effective wasn't hard, since patients often did recover in the days and weeks after being bled, blistered, purged, or gyrated. In the absence of controlled clinical trials, it was easy to interpret these recoveries as due to the psychiatrist's efforts and failures as indicating that the case was too serious or too deeply ingrained to respond to treatment. With their professional existence at stake, not for the first time, necessity became the mother of delusion and circularity the lifebuoy of the desperate.

PART FOUR

DO-IT-YOURSELF HEROES

DO-IT-YOURSELF HEROES

Philosophy [science] is such an impertinently litigious lady that a man had as good be engaged in law suits as have to do with her.

Isaac Newton in a letter to Edmond Halley, 20 June 1686.

A common theme runs through the preceding section and this one. Both concern individuals singled out as the father figures of their respective disciplines. With the exception of Pinel, the originality of our subjects has been greatly overstated. Nevertheless, these two sections differ in an important way. Weyer and Pinel played no part in the imaginative reconstruction of their lives that turned them into heroes. On the other hand, the evidence suggests that both of them were exceptionally compassionate individuals who deserve to be remembered for their personal qualities. The two scientists we'll examine in the following pages, however, were personally responsible for carving out their own giant reputations. The first, Sir Robert Watson-Watt, insisted on being regarded as the 'father of radar', the second, Selman Waksman, was recognized as the discoverer of streptomycin, the first antibiotic effective against the scourge of tuberculosis. Both men were remarkably successful in their campaigns of self-aggrandizement. Waksman became, in the public's mind, a pioneer who had opened up whole new possibilities for the treatment of disease. Watson-Watt was celebrated as a backroom genius who'd done more than perhaps any other individual to defeat the Luftwaffe in the year that Britain stood alone after the Fall of France.

Yet neither man had it entirely his own way. For a while at least both became mired in unseemly priority disputes with individuals who accused them of claiming too much. Such accusations are not, of course, uncommon in science. Examples spring quickly to mind from various centuries: the feud between Hooke and Newton over their respective roles in formulating the theory of universal gravitation; the even nastier spat between Leibnitz and Newton over the invention of calculus; the extraordinary dispute over who deserved credit for the isolation of insulin, be it

215

different members of a Toronto University team or a little-known Romanian; the controversy over who first identified the HIV virus, America's Robert Gallo or France's Luc Montagnier; the debate over why Jocelyn Bell didn't receive a share in the Nobel Prize for the discovery of pulsars; and, more recently, Raymond Damadian's claims, elaborated in a series of full-page advertisements in national daily newspapers, to have been deprived of his rightful share in the Nobel Prize awarded for the discovery of MRI technology. Nor are these kinds of debate confined to scientists. Motivated by patriotism, favouritism, or professional self-interest, those who write about the past also disagree about who should or should not be dubbed the father of one or other scientific field. Science remains, as Newton fumed, 'an impertinently litigious lady'.

But the cases of Watson-Watt and Waksman, like that of Jocelyn Bell, raise questions about the exercise of power by those in charge. An anthropologist spending time in dozens of scientific laboratories would encounter a wide diversity of management styles. Our anthropologist might also be struck by how each style reflects a different resolution of the same basic tension between the need, on the one hand, for effective direction and, on the other, the autonomy essential to the creative process. If we discount laboratories in which the dead hand of bureaucracy stultifies progress and those in which chaos reins, we can imagine a range of viable organizational options.

At one end are Type 1 laboratories. These resemble the private fiefdoms of enlightened despots. In such top-down labs, the head of department is responsible for all major decisions and closely instructs each research student. At the opposite end of the continuum are what might be termed Type 2, bottom-up labs in which graduate students have considerable personal autonomy and receive advice from above only as the need arises. When major discoveries are made by graduate students in the second type, natural justice requires that the head of department judges, case by case, a fair apportionment of credit. In many instances the student's name will appear first on any papers. Conversely, in more autocratic labs, the head garners most of the recognition. Not only are students using his or her methods, equipment, and resources, they've been told exactly what to do and they are directly building on their superior's prior achievements. In such cases, the head's name appears first on most publications.

Watson-Watt and Waksman both enjoyed Type 2 relationships with able subordinates. But in pursuit of higher individual acclaim, they later chose to redefine their relationships as having been more like Type 1. Watson-Watt was lucky in that his assistants kept their sense of grievance largely to themselves. Waksman was less fortunate. The man he sidelined, Albert Schatz, fought doggedly to have his own contributions properly recognized. Schatz took Waksman to court and won. But what happened next has much to say about the kinds of tension that can develop between junior researchers and their supervisors, for it wasn't Waksman's career that was ruined by the court case, but that of his subordinate. Our anthropologist could no doubt shed light on why this was so.

In *The Selfish Gene* Richard Dawkins noted that 'entire scientific reputations may have been built on the work of students and colleagues.' Cases of this extreme nature are almost certainly rare. But as cultural habits in the laboratory are seldom formally regulated, some bosses are accustomed to gaining marginally more credit than is really their due, as a perquisite of their status. This is implied by the fact that Waksman seems to have found it relatively easy to recruit high-level support in his battle against Albert Schatz. Due to a common interest among senior figures in keeping graduate students in what is deemed to be their proper place, nearly all leaders of the field angrily dismissed Schatz's claims to have been elbowed off the podium. The following two cases are in no sense representative of conduct in modern labs, but because of their extreme nature, they do usefully highlight some of the social dynamics that have been involved in the apportioning of rewards in science.

The first casualty of war   9

Sir Robert Watson-Watt (1892–1973)

I modestly believe myself to be the Father of Radar.
Sir Robert Watson-Watt,
Three Steps to Victory (1957).

The saying that 'history is written by the victors' has become a timeworn cliché; yet rather like Lord Acton's remark, 'Power tends to corrupt, and absolute power corrupts absolutely,' it conveys one of the few essential truths about the human condition. No recorded society seems to have resisted the urge to launder and romanticize its own past: concealing atrocities, rubbing out defeats, turning worldly careerists into selfless defenders of liberty and truth, and inventing heroic foundation myths. In the case of science, unlike politics and warfare, the victor's perspective is generally the one closest to the truth; however noble the lie, the scientific method should eventually flush it out. But the same cannot be said for the making of great scientific reputations. And where a hero's status is bound up with a patriotic struggle, historical truth is an early casualty.

This chapter explores how being on the side of the victors enabled a British scientist to garner almost the entire credit for one of the most important technological innovations of the twentieth century. In Britain at least, Sir Robert Watson-Watt is widely recognized as the sole inventor of radar, a technology that played a crucial part in the Battle of Britain, itself seen as one of the key episodes determining the outcome of World War II. By allowing Fighter Command's hard-pressed fighters to remain at rest until German bombers were approaching British airspace, it hugely increased the RAF's effectiveness. Fighters could be vectored into the attack with full fuel tanks, enabling them to inflict maximum damage on enemy bombers before and after they dropped their deadly payloads on British towns and cities.

LEFT: *Robert Watson-Watt © Bettmann/Corbis.*

Considering the vital importance of radar to the war effort, it's no wonder that Robert Watson-Watt emerged as one of the darlings of the postwar period. Among the 'boffins turned magicians' credited with helping defeat the Axis powers, he achieved signal respect. And Watson-Watt positively exulted in this heroic status. In his lavishly self-congratulatory autobiography, *Three Steps to Victory*, he repeatedly compared himself to the Duke of Wellington. Without the original's strong vein of self-deprecating humour, to summarize his contribution to radar's development Watson-Watt co-opted the Iron Duke's 'By God! I do not think it would have been done if I had not been there.' Later in the book he declared, 'I modestly believe myself to be the Father of Radar.'

A cynical reader of *Three Steps to Victory* might have wondered why this much-admired man had gone to such lengths to ensure that the words 'radar' and 'Watson-Watt' were linked in the public mind as surely as Columbus and America or Darwin and evolution. His relentless insistence that he alone had fathered radar, and his often cutting dismissal of rival claimants for the title, might have raised suspicions. But in a period in which Britain drew comfort from its wartime successes to offset the hardships it encountered in the postwar world, Watson-Watt was given free rein to write his own history. The difficulty is that while there's no doubt that he played a crucial role in giving the RAF an effective radar system in 1940, the story he told in *Three Steps to Victory* contained a number of serious distortions. Historians have now collected enough counter-evidence to justify a major overhaul of the way in which the radar story has usually been told.

The Chain Home system

Persuading the British establishment to take the idea of radar seriously was no easy task. In 1932 Stanley Baldwin had announced with naïve defeatism that 'the bomber will always get through.' It was a view that the pioneers of radar development would struggle to overturn. Fortunately, this was one of the many issues on which Winston Churchill and Baldwin disagreed. In 1934 Churchill helped form a special scientific committee, under Sir Henry Tizard, to investigate the possibilities of air defence. As planes were now travelling much too fast for sound waves to provide an early warning of a bombing raid, the use of infrared light was explored. This was soon shown to be a non-starter as well. Serious attention was next given to rumours that Adolf Hitler's scientists had invented a giant ray-gun that could vaporize entire cities. A concerned Air Ministry let it be known that anyone who could develop a ray capable of killing a sheep at 200 yards would win £1000.

Although nobody ever claimed the prize, Tizard's committee looked seriously into the possibility of devising a ray-gun that could focus electromagnetic radiation onto incoming planes to detonate their bombs, disable their engines, or even boil their pilots' blood. Watson-Watt, based at the National Physical Laboratory, was first called in to evaluate the death-ray scheme. He passed the job on to one of his assistants, Arnold F. Wilkins, who later recalled the memo as reading something like, 'Please calculate the amount of RF power which should be radiated to raise the temperature of 8 pints of water from 98°F to 105°F at a distance of 5 km and a height of 1km.' Realizing what was intended, Wilkins made the calculations, realised the ray's impracticality, and gave the idea short shrift.

It was at this stage that Watson-Watt suggested to Tizard's committee that resources be directed instead to researching the 'less unpromising' idea of using 'reflected radio waves' to determine the presence of enemy planes. Watson-Watt was well qualified to pursue this idea. He'd spent some time using radio waves to detect incoming thunderstorms. Developing a system to locate unseen aircraft was in many ways an extension of this earlier project. As a result, within a month of getting the committee's support, Watson-Watt was able to present a plan for transmitting radio waves and then analysing those reflected back to a receiver. Just twelve days later he had a workable system that was able to detect a Heyford bomber at a distance of eight miles as it flew over Daventry.

By August the following year, the nucleus of what was called the Chain Home (CH) network was up and running. Lattice towers 350 feet high were erected on the south and east coasts of England. The high-frequency radio pulses they transmitted were effective over a range of up to 150 miles, and each tower covered a 120-degree sector facing south into what would become enemy-occupied Europe. Pulses bouncing back from aircraft flying into a sector were picked up by receivers mounted on the towers. By August 1936 seven of these towers were in operation. Because they were well spaced along the coast, aircraft could be picked up by at least two towers and, using simple trigonometry, the staff of a central command centre could then calculate the aircraft's position.

Compared with what was to follow, the CH system was extremely low-tech. Indeed, the technology is often described as 'dead-end'. But it gave the British a twenty-minute warning of approaching aircraft during the Battle of Britain, usually long enough to enable fighters to be deployed to maximum effect. Like the Battle of Waterloo, the Battle of Britain was 'a damn close run thing', and it's almost certain that Britain would have lost the air war without the CH system. For this reason alone Watson-Watt deserved public admiration; but to him this wasn't enough. He was not going to rest satisfied with accolades for merely helping to implement a system that saved many thousands of lives and changed the course of the war. Watson-Watt wanted the kind of immortality that major scientific discoverers enjoy. But was he really entitled to this additional credit?

Germany's unsung pioneer

The invention of radar was the culmination of an international research effort arising most directly out of James Clerk Maxwell's 1850 prediction of the existence of electromagnetic waves and their subsequent discovery in 1887 by Heinrich Hertz. Within a decade, Guglielmo Marconi had introduced the first 'wireless'. And, soon after, the brilliant but unstable Nikola Tesla had speculated that the 'echoes' of radio waves could be used to locate distant objects. This was the fundamental concept of radar. Communicating by radio entails sending a radio signal to a distant receiver. Radar involves siting the receiver alongside the transmitter with the intention of picking up any signals bounced back. Radar was such an obvious extension of radio technology that the idea was almost bound to occur to some of the early pioneers of radio. The real difficulty lay in designing effective apparatus. A major breakthrough had been made in 1887 when Karl F. Braun invented the cathode-ray tube, which would later be used for presenting electronic beams upon a glass screen coated with fluorescent material. Reflecting on these developments, in 1958 Watson-Watt remarked that the invention of radar would have been technically feasible 'after a fashion, any time after about 1926'. In fact, even this assessment was off target by more than two decades.

Early in the morning on 10 May 1904, a crowd of reporters and passers-by gathered next to a bridge over the Rhine in Cologne. They were staring intently at two men who had set up a strange contraption on a platform under the middle of the bridge. When asked what they were doing, one of them, a talented young engineer called Christian Hülsmeyer, shouted back: 'This is an apparatus for preventing collisions at sea. The electric waves given out by this transmitter are reflected back to this receiver by a vessel, so that it can be detected at night or in fog.' By all accounts, the responses of the crowd varied from incomprehension to disbelief. Then a Rhine barge appeared in the distance and the apparatus was switched on. Almost at once, a bell on the contraption rang out loudly into the quiet of the morning. Hülsmeyer redirected the antennas into the sky and the bell stopped ringing. He then pointed them once more at the approaching barge. Again, the bell started to ring. The reporters knew that they'd witnessed something of genuine importance and rushed off to tell their editors. Next morning, news of Hülsmeyer's experiment appeared in dozens of Continental newspapers.

The 'Telemobiloscope' devised by Hülsmeyer was technically highly sophisticated, but the core of its design was combining transmitter and receiver in the same instrument. It meant that a captain on his bridge could detect approaching vessels from several miles away, even in conditions of heavy fog. The promise now appeared of an end to major maritime collisions. Unsurprisingly, Hülsmeyer anticipated that his patents would reap a rich financial harvest for him. So, in May 1904, he invited representatives from some of the period's shipping giants to see his Telemobiloscope in action. The second public trial was an emphatic success. A month later, the director of the Holland-Amerika-Line asked for a personal viewing. And on 9 June, he and dozens of other commercial shipping directors and engineers assembled in Rotterdam for the next demonstration. Loading his equipment on a small vessel, called the *Columbus,* Hülsmeyer directed the vessel up and down Rotterdam harbour as his Telemobiloscope unfailingly located any ship within a five-kilometre radius. His guests were certainly impressed. But to Hülsmeyer's consternation, no orders from commercial organizations transpired. The German Navy also turned him down flat. 'My people have better ideas!' Admiral von Tirpitz replied. Unfortunately for Hülsmeyer, shipping companies were not prepared to bear the cost of adopting an expensive new technology. Having already invested heavily in radio communication, most were content to rely on it alone. There was also some (probably unnecessary) concern that Hülsmeyer's new equipment might infringe the fiercely defended Marconi radio patents.

As a result, having spent a small fortune on registering patents, Hülsmeyer ran out of cash and was obliged to abandon his Telemobiloscope research. He eventually found a well-paid job in an entirely different field, and it was fifty years before he gained recognition. Eager for something to celebrate among the rubble and ruins of their postwar nation, in the early 1950s German politicians and scientists showered the elderly Hülsmeyer with praise. A street and a square in Düsseldorf were rechristened in his honour. And near his home town, an army barracks was named after him. But what probably pleased him most was that, at the International Radar Conference held in 1954, the German contingent introduced Hülsmeyer as the 'father of radar'. It was in precisely these emphatic terms that he was presented, in 1954, to Watson-Watt. The Briton seems to have treated the old man as if he were suffering from a harmless delusion.

Watson-Watt had come to the meeting forewarned, having been apprised of the claims being made on behalf of Hülsmeyer soon after the war. In 1942 Watson-Watt had been knighted for his contributions to radar development (at which time he gentrified his surname with a hyphen), but not long after the end of hostilities doubts had begun to surface. In 1949 Winston Churchill received a letter from one Annelise Hülsmeyer. In it she claimed that Watson-Watt was a poor second in the race to devise a workable radar system. Her father, she explained, had beaten him to the winning post by about thirty years. Although most letters of this kind can be dismissed as the work of jealous cranks, Annelise's had all the hallmarks of authenticity. It outlined the experiments, public trials, and initial press interest in the Telemobiloscope. It even cited the lapsed patent numbers registered in 1904 by her father in England. Radar, Annelise perhaps too flippantly concluded, was 'a German invention for the victory of the Allies!' Notwithstanding the strength of her case, she faced a daunting uphill struggle. After five years of bloody conflict, few Britons were in the mood to restore to their podiums forgotten German heroes, deserving or not. Accordingly, Churchill's reply was dourly formal.

When Annelise married soon after, her husband, Erich Hecker, also took up the cudgel, but he got no further. In February 1951, the British government simply advised him to take the matter up with a 'legal advisor'. With commendable determination, Hecker refused to give up. In April 1951 he tried a different tack and wrote to the British Patent Office. His 'obedient servant', A. V. King replied that no one was 'empowered to take any action'. King then used an argument of which Watson-Watt was to become a past master: defining the claims of other radar pioneers out of existence. According to King, Hülsmeyer could not be deemed the inventor of radar. 'In this country,' he explained, 'the term "Radar" is used to denote a system which includes means for determining the position of the distant object, that is, the distance thereof from the observer.' In short, unless an apparatus employing reflected radio waves accurately locates a ship or plane it cannot be considered radar. It was on exactly this basis that Watson-Watt later sought to eliminate Hülsmeyer from the radar hall of fame.

When he was writing *Three Steps to Victory*, Watson-Watt recognized that this German engineer couldn't be entirely omitted from his account. The mention of him was, however, of a very cursory nature. He wrote:

> Hülsmeyer of Düsseldorf made experiments on the reflection of radio waves in 1903, and in the following year patented a system intended to allow one ship to determine the direction in which another ship lay . . . No actual use of the device is known to me, although I met and talked with the inventor . . . in 1953 [actually, 1954].

Watson-Watt was claiming that Hülsmeyer's invention was an ineffective curiosity. He then dismissed the German's claim that radar had been invented in 1904, not in 1935, with lordly disdain: '[I have a] queer prejudice that one father is sufficient even for the lustiest of infants.' But we now know that both King and Watson-Watt were ignoring two awkward facts. First, as already described, Hülsmeyer's Telemobiloscope had been successfully tried out on at least three occasions. Second, not only did his system have a 'means of determining the position of the distant object', but according to the historian of radar, David Prichard, the method Hülsmeyer devised was of a sophistication that would not be matched again until the 1940s.

Watson-Watt's account of Hülsmeyer's apparatus was quite simply wrong. Whether or not he knew this is unclear. But it seems highly unlikely that by 1958 Watson-Watt had not at least heard of the Rotterdam Harbour trial. It would also be surprising had he not made some enquiries into the details of Hülsmeyer's claims before or after the two men met to discuss radar in 1954. On balance, it's probable that Watson-Watt marginalized the German's achievements because he realized the serious threat Hülsmeyer posed to his own reputation.

The radar pioneers

In the aftermath of war, when fresh evidence of Nazi atrocities was emerging with some regularity, trivializing a German radar pioneer required only the most rudimentary skill in the art of deception. But returning Hülsmeyer to the catacombs of obscurity did not end Robert Watson-Watt's problems. People closer to home (and on the right side throughout the war) also kept claiming that radar predated the CH system with which he'd been involved. Watson-Watt was to find it rather more difficult to cover over the traces left by these Allied radar scientists. Here much greater subtlety was required.

The Yorkshire-born physicist, Edward Victor Appleton, was probably Watson-Watt's closest British rival. In late 1924, the BBC had allowed Appleton to use its Bournemouth transmitter to fire radio waves into the sky. He'd then collected and analysed the radio echoes. By doing so,

Appleton demonstrated the existence of the region of the atmosphere known as the Heaviside Layer or ionosphere. In the process, he'd also made an important advance in improving the effectiveness of radar technology.

If a transmitter fires a continuous stream of radio energy at an unseen target the receiver is rapidly overloaded with identical echoes and the operator is unable to tell from how far away they're coming. If, on the other hand, the operator varies the wavelength of the transmitted radio waves, he or she can match the incoming with the outgoing signal. A simple calculation, based on the known speed of electromagnetic energy, then allows the operator to determine the target's range with a high degree of accuracy. This is precisely what Appleton did in 1924. By using waves of differing frequencies, he was able to establish that the ionosphere comprises a band between 90 and 150 kilometres from Earth's surface.

In 1929 Appleton next took his radar equipment to Norway to study the aurora borealis. While he was there he made two further innovations. The first was the use of 'spurts' of radio energy. Transmitting in short bursts, he waited until the echo arrived at the transmitter before firing off more radio waves. This made the measurement of distance between transmitter and target more accurate still. His second innovation was to develop a means by which the reflected signals could be represented using cathode-ray tubes. For this work Appleton won a Nobel Prize in 1947. Significantly, the ceremonial oration proclaimed that he'd first used 'radiolocation' in 1924. Watson-Watt himself had been closely involved in this work in a junior capacity, working under Appleton's direction. The CH system he later helped to devise was profoundly indebted to Appleton's imaginative use of radio technology.

Appleton, however, was no lone pioneer. By the 1920s American scientists were showing a strong interest in the idea of radiolocation. In 1922 Albert Hoyt Taylor and Leo C. Young had developed a primitive radar system that could detect a small steamer at several hundred yards' distance. Ten years later, they were assigned the task of investigating the potential for radio waves to be used in detecting enemy ships and planes. Like Appleton, they learned to use pulsed beams instead of a continuous wave pattern. And by December 1934, equipment they had developed proved capable of tracking a small plane as it flew up and down the

Potomac River. With the assistance of the Carnegie Institute's Gregory Breit and Merle A. Tuve, Taylor and Young also used their radar equipment to confirm Appleton's work on the size and height of the ionosphere. In 1938, after several naval trials, their radar system was fitted aboard the USS *New York*.

In Germany, radar was rediscovered in 1933 by Dr Rudolph Kühnold. With his guidance, the Gema company had a prototype radar available for trials in January 1934, and it was successfully demonstrated on 20 March in Kiel harbour. A few months later, Kühnold showed that his apparatus could locate ships as many as seven miles away. Like Taylor and Young, he too now switched to the more versatile method of transmitting pulsed radio waves. In September 1935, Kühnold's radar was exhibited to Admiral Raeder, the Naval Commander-in-Chief, who was impressed to see it spot ships twelve miles distant. Thus, at much the same time as Watson-Watt's prototype radar system was being judged a success, German physicists and engineers had devised their own radar technology.

After the war Watson-Watt did nothing to upset the popular British belief that the Germans had developed radar only after finding some intriguing apparatus on a British vessel captured in 1941. Indeed, myths seem to be an integral part of the radar story. During the war British intelligence sought to conceal the use of radar sets by their night fighters by putting out a story that RAF pilots were fed a diet of carrots to improve their night vision. For the same reason, the leading British night-fighter ace was known as 'Cats-eyes Cunningham'. Watson-Watt happily played up to these myths in his *Three Steps to Victory*. He knew perfectly well, however, that the Germans had devised several advanced radar systems well before the opening shots of World War II had been fired.

The fact that British intelligence felt the need to spread rumours about diets rich in Vitamin A indicates just how ill-informed they had been about German technical progress. In 1939 the job of finding out what the Germans knew was given to Dr (later Sir) R. V. Jones. At first Jones worked on his own from an office in MI6 headquarters. The British government had already received an anonymous package from a 'friendly German scientist' who had posted it through the letter-box of the British Naval Attaché in Oslo. It claimed that the Germans already had an operational radar system which, in September 1939, had enabled them to detect bombers approaching Wilhelmshafen, 120 kilometres from their target. As a result, sixteen out of twenty-four of the RAF planes were destroyed. At the time, the British had interpreted this as no more than extreme bad luck. The authorities were also quick to disparage the 'Oslo' report as 'disinformation.' It was an unfortunate mistake.

Soon after, in December 1939, radar equipment was found attached to the mast of the pocket battleship *Graf Spee*, which had been scuttled in the estuary of the River Plate. Once again the opportunity was lost. The report sent to London was filed away by an overburdened intelligence service and quickly forgotten. Only in 1940 did evidence that Germany had developed radar become irresistible. Photo reconnaissance brought to Jones' attention two oddly shaped circles in a field in northern France. Shortly before, one of Jones' growing team of assistants had taken a radio receiver to the south coast and monitored radio waves transmitted from across the Channel. Some of them intersected exactly with the circles photographed by the reconnaissance plane. On 24 February 1940, at a high-level meeting convened to discuss whether or not the Germans had radar, Jones presented his evidence. 'This was,' he wrote later, 'the end of disbelief in German radar.' A year later, an audacious commando raid on the French village of Bruneval enabled the British to capture a German radar unit and some of its operators. An accompanying radar specialist examined it on the return trip to Britain. He described the system used as a 'beautiful job'. Its sophistication made it clear that the Germans had been working on radar for about a decade.

To some extent, it's not surprising that the British were ignorant of German radar technology for so long. Instead of the vast towers erected by the British, the Germans favoured smaller, mobile units. Although their modest size made them less powerful than the CH system, it also meant that they were far less visible. This difference could have cost the RAF the Battle of Britain. Midway through the battle, the Germans switched their attack to the radar stations, having realized their significance. It was only Goering's foolhardy decision to switch from attacking radar emplacements to bombing London that gave Britain a reprieve. It should also be noted that the comparatively small size of the German equipment did not reflect technical inferiority. During the early years of the war the core of Nazi military strategy was an imperative to attack, so they produced highly effective range-finders for naval guns and mobile flak crews and deployed radio-based technology to improve the accuracy of bombing. Only in July 1940, with the menace of Allied bombing raids increasing, did the Germans recognize the value of a coordinated early-warning system. Subsequently, a highly effective system of 'air defence boxes' was set up along the coast of Europe, each equipped with radar, searchlights, flak guns, and fighter aircraft, which took a terrible toll on Allied bombers.

In comparison to these and later British and American efforts, the CH system can only be described as rudimentary. Watson-Watt's claim on our admiration should lie in his demonstration of the following very sound principle, 'in a crisis, a crude but operational system is infinitely preferable to a much better technology that might be developed given more time.' Perhaps this should have been called 'Watson-Watt's Law' to provide his proper epitaph. But as it would not have offered him entry to science's premier division, it wasn't an avenue he pursued. Instead, Watson-Watt set about an ultimately unsustainable rewriting of history.

'Genealogical embarrassments'

When Watson-Watt sat down to write *Three Steps to Victory* he could hardly ignore all these competing efforts. The best he could hope to achieve was to play down their importance so much that they quickly faded from memory. This is precisely what he strove to do. As such, out of a total of 450 pages, he devoted a mere sixteen to explaining the radar research that preceded his own.

His next step was to contort the definition of radar in such a way that the combined efforts of Appleton, Young, Taylor, Kühnold, and several others, became little more than the overture to his own grand symphony. As he candidly wrote, 'Genealogical embarrassments can only be avoided by resort to the philosopher's defensive, "It depends what you mean by radar." ' There followed several pages of hair-splitting pedantry in which he elaborated a definition of radar that deliberately conformed to anything that his radar system of 1935 could do and that its predecessors could not. He explained:

> No system which does not ensure the disentangling of
> information about a large number of widely and not-very-
> widely separated formations, each comprising a number of
> aircraft with varying separations . . . is, in my view, radar.

Appleton had said nothing about aircraft. Nor had he measured specific objects in the ionosphere. Taylor and Young had tracked single planes only, and their system did not allow them to do a great deal more. And by mid-1935, Kühnold's system had only been tried against ships, and not many of these. In each case, Watson-Watt invoked what was to become a trusty refrain, 'It was a bit of radar, but it was not radar.'

If one accepts Watson-Watt's definition, he would indeed deserve to be considered the inventor of radar. Yet it would be very unwise to do so, for it was an outrageously narrow and self-serving definition. It was tantamount to confusing the icing for the cake itself, and there is no reason why a different definition shouldn't be accepted in its place. Take, for example, the one devised for the Massachusetts Institute of Technology's *Principles of Radar*:

> Radar may be defined as the art of detecting by means of radio
> echoes the presence of objects, determining their direction and
> range, recognising their character and employing the data thus
> obtained in the performance of military, naval or other
> operations.

In April 1935, the CH system couldn't have told a Zeppelin from a biplane. And Watson-Watt lived in dread of the Germans realizing that, simply by attaching small lengths of wire to a few thousand balloons released in the floodlit area, they could effectively paralyse the entire radar network. In other words, were the MIT definition to be endorsed, the accolades would go instead to the mid-war physicists and engineers who developed methods of distinguishing not only between planes and metal strips, but between fighters and bombers as well. (According to Sir R. V. Jones, Watson-Watt was so sensitive to his system's vulnerability to jamming by the enemy's use of wires or aluminium strips that he opposed their use in the defence of Allied aircraft even when the air war had moved decisively against Germany. As it happens, the Germans had already come up with the same idea but also held it in reserve to protect the effectiveness of their own radar systems. When the Allies did deploy what was code-named 'Windows', it gave their bombers a major, albeit brief, respite from German flak batteries and night fighters.)

A much more reasonable definition of radar was provided by Watson-Watt himself in an encyclopaedia article of 1955. Perhaps because he had suppressed any mention of Taylor, Young, and Kühnold in this abbreviated history, and had mentioned Appleton, Breit, and Tuve in only the most perfunctory fashion, he was willing to present a more inclusive definition. 'Radar,' he wrote, 'is a method of obtaining information about objects at a distance by means of electromagnetic waves reflected from these objects.' Prosaic but accurate, it was a description Watson-Watt hastily abandoned whenever his paternity status came under attack.

Seeking royal recognition

After the war, the government set up a Royal Commission to consider which wartime discoveries warranted financial rewards from the state. Dozens of scientists, including Frank Whittle and Barnes Wallace, received large and tax-free cash payments. Watson-Watt, to his profound chagrin, was not included among this stellar company of scientific war-winners, and he felt badly used. His first appeal to the authorities was, however, coldly rebuffed by E. L. Pickles, the Patents and Awards Officer of the Ministry of Supply. Pickles seems to have felt that the knighthood Watson-Watt had received in 1942 was ample compensation for his labours. In any case, he simply couldn't see that Watson-Watt had a legitimate claim. 'We deny invention,' was his succinct reply.

In a frenzy of wounded vanity, Watson-Watt applied for a formal hearing and outlined at length his alleged entitlement to financial recognition. 'I request that the Ministry,' he wrote, 'should consider the granting of an ex gratia award in respect of inventions and designs of exceptional utility made by me in the field of [radar research].' In his choice of pronouns, he paid less than generous tribute to the rest of the team that had worked on the CH system, in particular his closest assistants Arnold F. Wilkins and Edward George Bowen. But realizing that he needed allies, Watson-Watt also managed to persuade his ex-colleagues to join him. Finally, in 1951, a Royal Commission met to debate his petition. It was said of Robert Watson-Watt that words 'bubbled from him like a fountain'. And, as a foretaste of what was to come, much of the first two days was taken up by his opening speech, in which he first outlined his unusual definition of radar. In all, his first of many orations took six hours to deliver.

As an acknowledged expert in the area of radio waves, E. V. Appleton was called in to help assess Watson-Watt's case. The committee transcripts make clear his antipathy towards the younger man. And it was presumably Appleton who inspired the first line of questioning: whether or not radar had been invented during the course of the ionosphere investigations of the 1920s, long before the detection of the Heyford bomber flying over Daventry. Ignorant of Hülsmeyer's much earlier achievements, Appleton presumably felt that in terms of both theory and technology, his entitlement to be recognized as the inventor was a very strong one. He doesn't seem to have been inclined to pursue his claim, but he was loath to allow the credit to be usurped by another.

Appleton's line of argument was fair, but this isn't how Watson-Watt saw it. The ionosphere work didn't employ true radar, the latter explained, for the simple reason that no attempt had been made by Appleton's meteorologists to 'fix by range and directional measurement, used together, the position in space of a small isolated patch of air'. In other words, they had not attempted to do for the atmosphere what he'd achieved with planes. Ergo, Appleton was not a radar pioneer. It was a line of reasoning that, as we saw a moment ago, confused a specific application with the development of a principle. The principle itself was, in any case, impossible to affix to a particular time, place, or person. This was, however, the best line of argument Watson-Watt had.

The Commission also tried to establish the extent to which Watson-Watt was personally responsible for the construction of the CH system. Once more Appleton was in the thick of the action, this time arguing on behalf of others. He told the committee that, having spent some time with the radar team in the years before the war, he was convinced that Bowen and Wilkins had been 'fundamental to the whole development'. When it came to delineating Watson-Watt's role, he portrayed him as a facilitator rather than an inventor. Watson-Watt, he explained, was always busy 'arguing and advocating with great eloquence, seeing that everybody got everything they wanted, writing memoranda ... and pushing people'. Given that Watson-Watt was seeking the acclaim of a scientific genius, this was faint praise indeed. Appleton conjured up an image of a busy administrator acting as a mother duck to a clutch of boffins. Watson-Watt's role was being made to seem akin to that of General Leslie Groves in the Manhattan Project. Groves was in charge of making sure the material resources were always in place so that J. Robert Oppenheimer, Ernest Lawrence, Enrico Fermi, and all the other physicists and engineers, could get on with the cutting-edge scientific research.

Watson-Watt was clearly horrified by Appleton's summary and the personal animosity it betrayed. In particular, he was offended by Appleton's affectionate references to Wilkins and Bowen as 'his' men, a category in which Watson-Watt, despite having worked with Appleton for several years, was clearly not included. Responding to Appleton's description of his role and the implication that most of the ideas came from his colleagues, Watson-Watt gritted his teeth and asked him to clarify. 'You have given a catalogue of the qualities which I put into the work,' he said, 'but did not explicitly include ... the scientific and technical work.' Appleton's reply was judicious: 'It is for the court to find out how much of the ideas which you are putting forward arose from their contribution.' Even in the initial memorandum with which the British military's development of radar began, Appleton was implying, Watson-Watt's contribution was impossible to disentangle from those of his immediate colleagues.

Remembering Wilkins

Appleton clearly felt that one colleague in particular had been deprived of the esteem he deserved: Arnold Wilkins. He had none of the flair for self-aggrandizement displayed by Watson-Watt; he seems to have simply got on with the job, leaving the search for fame and glory to his superior. But in 1977 he deposited in the Churchill Archives at Cambridge University an account of the development of the CH radar system in which he was rather less than complimentary about his by then deceased boss. Perhaps emboldened by age and time, Wilkins asserted that the idea of using radio waves to detect approaching aircraft had originated with himself. From what we know of the characters of those involved and the opinions of several individuals closely associated with the development of British radar, we can accept Wilkins' claim with some confidence.

For several years prior to being assigned to Watson-Watt's team, Wilkins had been working in Slough at the Radio Research Station. In 1931 he'd conducted some research using Post Office radio equipment in Colney Heath, near St Albans. During this research stint he was told that transmissions frequently suffered interference when aircraft from the nearby de Havilland's aerodrome flew overhead. The radio operators deemed this a mere 'nuisance', but Wilkins put it at the back of his mind. He recalled it in 1935 when asked to evaluate the Tizard committee's death-ray scheme. Once it became clear that this was not viable, he thought instead of what he'd learned at Colney Heath. He then told Watson-Watt that the disruption in radio signals might offer a viable means of detecting aircraft. This, Wilkins asserted, then led on to Watson-Watt's famous memo to Tizard's committee. The idea of military radar, Wilkins wrote in his memoir, did not come to him as a flash of brilliant insight, but as an 'obvious thing to consider'.

It cannot be denied that Watson-Watt deserves high praise for immediately recognizing the importance of this idea and driving through its application in time for the Battle of Britain. But even at the height of the war, it seems that he was laying the ground to claim a great deal more than was his due. Wilkins describes how Watson-Watt first began to assert his primacy in radar research. 'When the existence of RDF [radar] was disclosed to the public in 1942 and Watson-Watt was disclosed to be the inventor,' Wilkins wrote, 'he told me that he had thought of using radio waves for aircraft location before 1935 . . . and had put his ideas to Dr. W. S. Tucker.' This was news to Wilkins. His recollection was that when he'd first proposed the notion of military radar to Watson-Watt, his usually loquacious boss had made absolutely no reference to having had his own thoughts on the subject. Wilkins discreetly investigated Watson-Watt's claim. He was not surprised when he discovered that 'Tucker had no recollection whatever of it.'

Wilkins, however, didn't put up much resistance, and this presumably encouraged Watson-Watt to press on with his campaign. His next task was to secure some kind of legal hold over the development of the CH system. His approach was crafty. Wilkins relates: 'Watson-Watt told Bowen and me that the Air Ministry had advised him to take out a secret patent for RDF. This, he thought, was rather a nuisance but he would have to comply with the wishes of the A[ir] M[inistry].' In fact, there's no evidence that he was put under any kind of pressure to do so by the authorities. Unlike the invention of radar, the idea of registering a patent was almost certainly an idea that originated from Watson-Watt. Wilkins continued:

> He went on to ask us whether we wanted to be named as co-inventors on the patent and there was something in the way he asked the question that gave me a strong feeling that he wanted to keep us out of it. If we had been considering a scientific publication rather than a patent, I would have felt insulted to have had no acknowledgement of joint authorship and would have pressed to be included . . . As we were merely considering a patent I told Watson-Watt I would not wish to be included in the authorship. Bowen, rather reluctantly, decided not to press his claims in the matter.

Watson-Watt probably couldn't believe his luck when Wilkins and Bowen agreed to cooperate with his scheme. But their willingness to do so is perhaps not surprising.

After the war, Appleton and others severely criticized Watson-Watt for having spent time registering patents during a period of wartime emergency. When members of the armed services were fighting and dying across the globe, when merchant seamen were perishing by the thousand in shipping convoys, and, on the home front, civilians too were in the firing line, taking time out to protect intellectual property rights seemed self-centred to say the least. In the fraught year of 1942, it may well have been that the minds of Wilkins and Bowen were too focused on the fight against Nazism to be much distracted by patent applications or with following up their suspicions about Watson-Watt's motives. This offered Watson-Watt an irresistible opportunity.

His assertions were all the more reprehensible because the

development of Wilkins' initial idea had required the cooperative efforts of dozens of scientists. None the less, the name of Wilkins still stands out. We now know that he performed a vital role in designing the radar system used in 1935 to detect the Heyford bomber passing over Daventry. Typically, on the night before the trial, he worked late into the night by candlelight, making last-minute adjustments in the back of a van. Wilkins, Bowen, and another scientist, L. H. Bainbridge-Bell, also played decisive parts (alongside Watson-Watt) in the subsequent setting-up of the CH system. The moral of their part in this story seems to be that, in science as elsewhere, fame is rarely the reward of modesty.

During the proceedings of the 1951 Royal Commission, Wilkins said little to embarrass his old boss. But enough had been disclosed by others, Appleton in particular, to persuade the Commission that Watson-Watt's words were not to be trusted. Seemingly unperturbed by the proceedings, Watson-Watt concluded his case with an exposition a third of a million words long that took four days to deliver. Like a general bereft of new ideas, he seemed to be resorting to a strategy of attrition. And, to the surprise of many, Watson-Watt carried the day.

But it wasn't the overwhelming victory for which he'd hoped. In summing up the reasons for the Commission's eventual decision, Sir Henry Tizard explained:

> I recognise the value of a new idea to do something that has
> never been done before. But what really matters, in my opinion,
> is to have the ability, the good judgement, the energy and
> powers of organisation, to bring the idea to a practical fruition.
> All these essential qualities were provided by Sir Robert
> Watson-Watt and his colleagues.

This is a masterpiece of British compromise. In effect, Tizard's statement might be summarized as saying: 'Of course Appleton's right, but in the interests of a quiet life and not tarnishing the image of a man the public has been taught to consider a world-class hero, we'll give him the money anyway.' Watson-Watt went away apparently satisfied, but the British government could content itself that the final summing-up had hardly budged from E. L. Pickles' initial denial of invention.

The 'father figure'

In bringing this story to a close it's instructive to reflect on our general tendency to attribute each major conceptual or technological development to a single father figure. Watson-Watt used this paternity metaphor to great effect, but as a model of how scientific progress usually occurs it can be extremely misleading. This is because, except in rare instances, developments in science can't be carved up into discrete segments, each allotted to a single individual. Every scientific success is the culmination of a research effort involving numerous individuals over long periods of time.

For all its undoubted originality, even Charles Darwin's evolutionary theory drew upon a wide range of pre-existing concepts and observations: from the fields of zoology, botany, geology and animal husbandry, as well as political economy. And, even by his own admission, Isaac Newton stood upon the shoulders of men like Descartes, Kepler, Galileo, and Hooke. Watson-Watt's achievement required rather less creativity. The CH system could never have come into being were it not for the Victorian physicists who formulated the electromagnetic paradigm, the many scientists, including Hertz, Marconi, and Tesla, who pioneered the investigation of radio waves, Appleton's use of radar technology for investigating the atmosphere, not to mention all the work of Watson-Watt's subordinates. Given this host of contributors, on what basis should one or other of these individuals be dubbed the true father of radar? How can one affix this label without making an invidious, not to say arbitrary, choice?

No achievement in science is exclusively the product of one brain. We can still, of course, point to instances where everything 'came together' in the mind of a particular individual. Yet except in a few cases, it's not obvious that this individual deserves greater credit than many who went before. Where a scientist forged a bold new synthesis and added important new insights, as with Newton, Darwin, and Einstein, one can easily justify singling them out. But it's more typical for the individual celebrated as a father figure to have achieved the scientific equivalent of running the last leg of a relay race. It's one of the oddities of scientific culture that this person often stands alone on the podium.

The myth of the 'father figure' distorts reality in other ways. Some ideas do begin as imaginative leaps performed by outstanding individuals. But if it has any promise at all, such an idea will soon attract the attention of others, and its development nearly always results in a competitive battle to advance the new field. So it proved with radar.

From the pioneering work of Faraday, Maxwell, Hertz, and Marconi, by the 1930s there were teams working on the problem in Britain, America, Germany, France, Japan, and Czechoslovakia. Nor were they all pursuing radio research in their own vacuums. Whether in the form of published articles or the spoken word, most teams were at least partially aware of what their competitors were up to. Watson-Watt, for instance, built directly on the research of Appleton and several other British pioneers. He chose to present the contributions of his predecessors as mere tributaries that flowed into the mighty river of his own achievements, but the reality was very different. As we've seen, quite apart from the fact that other teams in Germany, America, and Britain had employed radar long before Watson-Watt, only months after he successfully tried out his system on the skies over Daventry, Kühnold in Germany and Young and Taylor in America had developed their own military radar systems. It's another curiosity of the way recognition is apportioned in science that there are no rewards for coming second, even though, as the physicist François Arago pointed out, 'questions as to priority may depend on weeks, on days, on hours, on minutes.'

If the metaphor of the father figure overstates the originality of individuals, it also exaggerates their prescience. New facts about nature aren't revealed in the same all-or-nothing way in which, say, Columbus discovered the Americas. Scientific theories are assembled piecemeal and take time to acquire credibility. Moreover, in the process of verification theories seldom remain in their original form. This unquestionably applies to Watson-Watt, for even if he did manage to get an effective defensive radar system up and running before anyone else, it was soon eclipsed in sophistication by those developed in Germany, America, and in Britain. The radar technologies deployed by the Allies after the Battle of Britain owed only a modest amount to Watson-Watt's efforts.

But in staking his claim as the inventor of radar, the bogus paternity metaphor suited Watson-Watt well because it so effectively masks many of the more typical features of the scientific enterprise. In particular, it allowed him to acknowledge the efforts of his predecessors without much danger of their being given a direct role. 'I am prepared to recognise grand-parents and great-parents in geometric progression,' he remarked with faux generosity. There's room in any genealogy for plenty of relations, but only one father. So Appleton, Young, Taylor, Kühnold, and the rest, became members of the extended family, consigned to the background of radar's family portrait. Considering how much the work of such men contributed to the development of the systems with which Watson-Watt's name is indelibly associated, and the fact that rudimentary forms of radar had been used in 1905, 1925, 1926, and 1929, it required enormous effrontery on Watson-Watt's part to invoke this metaphor. Further, given that radar was really a development of the radio technologies devised through the labours of Maxwell, Hertz, Marconi, and others, it was hardly just to award them, at most, one-sixteenth of the overall credit.

But perhaps the most insulting extension of Watson-Watt's paternity model was his assigning the role of 'mother of radar' to Arnold Wilkins. At the beginning of the twenty-first century this might sound like an act of inclusive generosity. Yet when the term was coined, it most certainly was not. In the patriarchal world of postwar science, 'mothers' were very definitely second-class citizens. Accordingly, the contribution Watson-Watt ascribed to Wilkins in his maternal role was that of giving the 'embryonic war-winner a good start in life'. 'She' provided selfless succour, not fecund inspiration. Wilkins' reputation for self-effacing loyalty was well earned. In 1951 he dutifully accepted Watson-Watt's persistent claim that he was the originator of the Tizard Committee memorandum and what flowed from it. But as the archives of Churchill College in Cambridge suggest, the truth was somewhat different.

Selman A. Waksman (1888–1973)

> A New Jersey farmer was upset: his chickens were
> catching a strange infection from barnyard dirt. He
> took the birds to the Rutgers University laboratory of
> microbiologist Selman Waksman, who analysed the
> barnyard soil and isolated the problem—a peculiar
> fungus. In the process, Waksman fortuitously
> discovered that the . . . fungus produced a chemical
> agent that slowed the growth of certain bacteria.
> Ken Chowder, in *The Smithsonian* (volume 23, 1992).

On 7 November 1949 a scientist called Selman Waksman appeared on the front cover of America's *Time* magazine. Neither before nor since has a soil microbiologist been so honoured, but in 1949 his story made nearly perfect copy. As *Time*'s leading article explained, Waksman was responsible for a medical triumph of such life-saving magnitude that only the development of penicillin could rival it. His laboratory had first come to the public's attention five years earlier when it announced the discovery of a substance, dubbed streptomycin, that cured tuberculosis. Although this disease is now staging a serious comeback, we in the West now tend to associate it more with the romantic fatalities of operatic heroines than with real-life tragedy. But from the times of the pharaohs, when its symptoms were first recorded, to the mid-1940s, tuberculosis had destroyed an estimated two billion lives. As late as 1937 it was still the leading cause of death in the United States, and tens of thousands of people were confined to rural sanitoriums. Waksman's public announcement in 1945 seemed to herald the end of one of the most dangerous bacteria with which humans have had to contend.

The onset of the most common form of tuberculosis was marked by sudden weight loss and severe respiratory problems, followed by

LEFT: *Selman A. Waksman.*

the coughing up of bloody sputum. Recoveries were few, and death painful and protracted, so the discovery of a drug effective against this destroyer of lives was always going to make headlines. But the story had another powerful dimension. Waksman had been born into a poor Jewish family living in a small market town just outside Kiev in Russia. As a Jew under the tsars, he was denied access to a full secondary education. Not prepared to have his life so circumscribed, Waksman had emigrated to the United States in his early twenties. His abilities were such that within a year he'd secured a state scholarship to enter Rutgers College in New Jersey. From there he graduated in 1915 with a degree in agriculture. Having obtained a Ph.D. degree at the University of California, he was then invited back to Rutgers, where he spent the rest of his career. By the time he retired, Waksman was a Nobel Laureate and the Director of Rutgers' internationally famous Institute of Microbiology.

Starting out with little aside from talent, courage, and a patchy education, Waksman had risen to become one of the best-known scientists in the world. It was one of those extraordinary stories of triumph over adversity so beloved of newspaper editors. In the United States it had the added bonus of mapping perfectly onto the ideal of the American Dream. There was more. For the miracle drug Waksman produced did not originate in some hi-tech laboratory, but had been found in a sample of farmyard soil. In this too, Waksman's adoptive nation could take pride. With a true frontier spirit, he hadn't been afraid to get his hands dirty. Unpretentiously, he'd followed the American ethic of hard work and enterprise. His reward had been success on a global scale. In 1952, Waksman travelled to Oslo to collect the Nobel Prize for Physiology and Medicine. And the audience at the award ceremony heard of how streptomycin had already saved the lives of many thousands of tuberculosis sufferers. 'As physicians,' the host enthused, 'we regard you as one of the greatest benefactors to mankind.' But this inspirational story has a dark underside.

Back in the US, there were those who thought from the outset that the Nobel Prize was being awarded to the wrong man. They believed that the 'real discoverer of streptomycin' was not Waksman but one of his assistants: a man named Albert Schatz. He too was the child of

Russian Jews, his grandparents having fled the anti-Semitism of tsarist Russia several decades before Waksman. As a result, Schatz had been born in Connecticut, where his family scraped a bare living from a small Christmas-tree farm. The acrimonious dispute between these two men, Waksman a seasoned and well-respected researcher, and Schatz an ambitious and idealistic graduate student, sullied the triumph of the advent of streptomycin and all but wrecked Schatz's career.

Those who have recently looked into the case accept that Schatz had much to feel bitter about. As will be seen, having initially recognized him as the official 'co-discoverer' of streptomycin, Waksman then worked assiduously to play down his junior colleague's contribution to what became arguably the proudest achievement of American medical science. We're all aware of instances in which rival scientists have engaged in bitter disputes for priority over this or that discovery. The intense feud between Schatz versus Waksman is a classic case. But its particular significance lies in the vexed question of a scientific supervisor's 'ownership' of work actually carried out by a research student.

Humble beginnings

Waksman's first appointment to Rutgers had been as a soil bacteriologist. By 1940 he was heading the university's new Department of Microbiology and specializing in the study of the kinds of soil micro-organism that kill other microbes. But his initial interest in microbiology was agricultural rather than medical. It was one of his former students, René Dubos, a researcher at the Rockefeller Institute in New York, who shifted the focus of microbiological research to the hunt for antibiotics. In the early 1930s this Frenchman made the very important discovery that the little-known 'cranberry bog bacillus' produces an enzyme that can pierce the thick outer membranes of pneumococcus bacteria, leading to their destruction. Having isolated and purified the enzyme, Dubos demonstrated that it could cure otherwise lethal pneumococcal infections in laboratory animals.

Waksman soon heard of Dubos' success and quickly realized that some of the many species of soil micro-organisms he'd discovered might also inhibit the activities of lethal bacteria in the living body. He first chose to investigate the actinomycetes, a group of organisms with characteristics of both bacteria and moulds. In the late 1930s Waksman and his team of assistants and graduate students mounted a large-scale screening programme. Hundreds of soil types were collected and their resident actinomycetes isolated. These were next cultivated in agar jelly on glass slides or Petri dishes, and different species of bacteria streaked across them. The researchers then waited until the bacteria had had a chance to grow. If they failed to do so in the region of an actinomycete colony, then he or she knew that that particular actinomycete had antibiotic properties. This was not especially hi-tech science, but it did require technical skill, patience, and a dogged willingness to continue hunting despite week after week of barren effort.

Several months into the research, Waksman and his team had their first taste of success when they isolated two compounds, actinomycin and streptothricin, with striking antibiotic potency. Unfortunately, as would often prove to be the case, it was quickly shown that both were far too toxic to be used medicinally, so the search continued. It was during the early phase of this screening programme that Albert Schatz joined Waksman's team. When the United States went to war in late 1941, Schatz was posted to the Medical Department of the Air Force in Florida. But, in his spare time, he searched for new moulds and actinomycetes in the Floridian soils, swamps, and coastal waters. Anything he considered interesting, Private Schatz dutifully sent to Waksman for further analysis. In return, Waksman kept him abreast of advances in the field of microbiology.

Discharged from the army with a back ailment in 1943, Schatz returned to Rutgers, and Waksman agreed to take him on as a doctoral student assigned to look for new soil-based antibiotics. It's important to appreciate from the start that Schatz was no ordinary subordinate. Having returned to Rutgers as a mature student with an exemplary academic record and plenty of research experience, he was far more capable and much less dependent than most of Waksman's other graduate students. In particular, Schatz was perfectly content to work on his own initiative without regular supervision.

The breakthrough

Shortly after Schatz's return to the lab, it was suggested to Waksman that he include the bacteria that cause tuberculosis among the organisms streaked across his plates of actinomycete colonies. At the time there was some scepticism about the usefulness of doing so. This was largely because the germ that causes tuberculosis, *Mycobacterium tuberculosis*, has evolved a tough, waxy coat that makes it very hard to destroy. Nevertheless, Waksman agreed to the proposal. He also accepted Schatz's offer of doing the tests as part of his research.

Schatz began with the harmless form *Mycobacterium phlei*. Soon after, he recognized that an antibiotic effective against this mild strain might not work against the lethal form. He therefore switched to a particularly virulent strain of *M. tuberculosis* known as HR 37. Fearful lest any of these germs escape, Waksman promptly moved Schatz into a lab located in the basement and got him to promise never to bring any tuberculosis specimens up to Waksman's own third-floor laboratory. That Schatz's blood serum was later shown to contain antibodies against tuberculosis indicates just how brave or foolhardy he was being. And, in shifting his student to the basement, Waksman was only exercising proper caution. But his concern for Schatz's welfare was not as fatherly as it might have been, for his laboratory was equipped with few of the precautions, such as ultraviolet lighting and air filters, routinely installed in other laboratories researching deadly forms of tuberculosis. Whatever the reasons for this, it's clear that Schatz was taking a serious gamble with his life.

It turned out to have been worth the risk. A couple of months into his screening programme, on 19 October 1943, Schatz isolated two strains of an actinomycete that did something extraordinary: it managed to prevent the usually remorseless tuberculosis germs from multiplying. One of these strains had been isolated by Schatz from a heavily manured soil he'd had delivered to his laboratory. The other had been obtained from the throat of a healthy chicken and given to him by a graduate student working for the Rutgers poultry pathologist. Soon after, both strains were identified by Waksman as members of a species of actinomycete he'd first found in 1916. The organism was dubbed *Streptomyces griseus* and its product streptomycin.

In early 1944 the discovery was announced in a scientific journal. The order of names in a journal article is supposed to denote the relative contribution made by each of those involved. In this critical paper Schatz's name appeared before those of Elizabeth Bugie (another graduate student) and Waksman himself, implying that Schatz had done the lion's share of the work. Soon after, another paper appeared, again with Schatz as senior author, detailing streptomycin's effect on tuberculosis germs. Neither paper said much about the compound's potential therapeutic uses, but this was no more than sensible caution. Curative panaceas for tuberculosis had been announced before, only to result in bitter disappointment. Waksman's team knew they would not have a wonder drug until they had shown that it worked in living bodies without having overly toxic effects.

So Waksman arranged some preliminary tests. He procured a batch of fertilized chicken eggs and split them into two groups. Into the first set his researchers syringed lethal bacteria plus a good dose of streptomycin. The remaining eggs, designated as controls, were injected with the harmful bacteria alone. A few days later Waksman and his colleagues were thrilled to find that while the second set of embryos had died and withered away, those eggs treated with streptomycin had hatched chicks in perfect health. With every expectation of success, they now made contact with teams of medical researchers who could perform much more decisive clinical tests.

The first trials

William Feldman and H. C. Hinshaw of the Mayo Clinic, Minnesota, quickly stepped in and agreed to test streptomycin on guinea-pigs infected with tuberculosis. If Schatz thought that, having made the discovery, he could sit back and relax, he was very much mistaken. In the following weeks he worked gruelling hours in order to produce enough streptomycin for Feldman and Hinshaw to carry out their all-important test. Schatz later recalled how he slept on the laboratory bench so he could keep his distilling apparatus in operation all through the night. To prevent the liquid in his distillation flasks from boiling away, he drew a red line across them and asked the night porter to wake him whenever the liquid level fell below it. This way he maximized the production of streptomycin and provided Feldman and Hinshaw with enough for them to run a small, but very encouraging, laboratory trial.

So impressed were they by the results that they implored the New Jersey pharmaceutical giant Merck to begin production on an industrial scale. As war was still raging, Merck argued that its priorities lay with producing penicillin, an antibiotic effective against the sorts of bacteria causing deadly infections in wounds. But Feldman and Hinshaw reminded the Merck bosses that more people died of infectious disease after World War I than had been killed during it. It was therefore imperative, they argued, that massive supplies of streptomycin be manufactured to help avoid a repetition. Merck eventually agreed.

On 15 November 1945, Merck's streptomycin was finally tried out on a human being. A 21-year-old patient at the Mineral Springs Sanatorium in Canon Falls, Minnesota, who was dying from pulmonary tuberculosis, agreed to the treatment. It was a complete success. Lots more cures were effected over the coming months until Feldman and Hinshaw found themselves deluged with begging letters from all around the world. A black-market trade quickly developed, with some unfortunates being sold baking powder passed off as streptomycin. Waksman found himself the man of the moment. Invited on dozens of trips abroad, he was sometimes driven to a hospital and asked to administer streptomycin to the tuberculous children of senior officials in his host countries.

At this stage, no doubt, every member of the team was suffused with pride. For Waksman it was the crowning moment of a research project he'd pioneered. For Schatz it represented the finest possible start to his career as a microbiologist. On the very best of terms, the two men appended their signatures to a patent assignment recognizing them as 'co-discoverers'. They also shook hands on a gentleman's agreement that neither would derive any financial gain from the discovery. The worm, however, was already in the apple.

In making this promise Waksman was being disingenuous to say the least. Unknown to Schatz, he'd already brokered a deal with the Rutgers Research and Endowment Foundation, giving him 20 per cent of all the streptomycin royalties. With demand for the drug fast outstripping supply, this covert arrangement was worth a fortune. And even though

Waksman ploughed most of the money back into the building of a research centre (later named after him), from the outset he'd been less than honest with his junior colleague.

The aftermath

Meanwhile, Schatz had written up his work on streptomycin, been awarded his doctorate, and had taken up a position at the Hopkins Marine Station, Pacific Grove, in California. On the surface his relations with Waksman, now limited to occasional letters, continued to be amicable. On three occasions Schatz received packages from Waksman containing official documents requiring his signature and a cheque for 500 dollars. He was somewhat nonplussed by the financial gifts, but Waksman presented himself as a generous benefactor and Schatz needed the money, so he didn't press the point. Then Schatz heard via the always busy academic grapevine that Waksman had reneged on their agreement and was cashing in on the discovery of streptomycin. This did not come as a complete shock to him.

It seems that doubts had slowly been growing in Schatz's mind as to the probity of Waksman's conduct. During his last weeks at Rutgers he'd begun to feel that his boss had been writing him out of one of the greatest episodes in the history of medicine. Reading about the discovery in newspapers and magazines, Schatz hardly ever saw his name mentioned. And, as Waksman was festooned with awards and media attention, Schatz was still struggling to get his career off the ground. So, in a tersely worded letter, written in early 1949, Schatz asked Waksman several pointed questions. Waksman's haughty reply did nothing to allay his suspicions. Having acknowledged Schatz's importance in a few areas, Waksman implied that he thought his ex-student was behaving like a greedy child. Your contribution, he wrote, comprised 'only a very small part of the picture in the development of streptomycin as a whole'. Far from being overawed by Waksman's overbearing, dismissive tone, Schatz instituted legal proceedings.

The litigation dragged on through most of 1950, causing Waksman, Rutgers, and Schatz such acute embarrassment that, just after Christmas, all parties called for the case to be settled by the judge out of court. The judge's decision went a considerable way towards vindicating Albert Schatz. He reduced Waksman's 20 per cent share of the royalties to 10 per cent. And of the remaining 10 per cent, 3 per cent went to Schatz, and the remainder to all those who had played any kind of role in the discovery, including Mr Adams, the laboratory dishwasher. Given the facts that have now emerged, we might think that the judge's settlement erred on the cautious side. None the less it was a signal defeat for Professor Waksman.

At the time, however, few were inclined to congratulate Schatz. Most of Waksman's colleagues and students expressed disgust that this graduate student had taken the matter to court and thereby seriously embarrassed one of America's leading scientists. As the scientific community closed ranks around Waksman, even those who agreed with the verdict tended to see Schatz as the kind of bolshy, litigious character they could well do without.

When he heard that Waksman had been awarded a Nobel Prize as the 'discoverer of streptomycin', Schatz was working in a small agricultural college in Pennsylvania, effectively out of contention for the best jobs. When the principal of the college wrote to the Nobel Prize Committee to protest, the president and secretary wrote back saying they'd never even heard of Albert Schatz. In despair, Schatz wrote first to the King of Sweden, Gustav VI, and then to all the leaders of the field, Alexander Fleming, Howard Florey, A. B. Sabin, and Hans Krebs, asking for support. None replied to Schatz and only Sabin made any kind of formal response. He wrote to Waksman expressing his horror that a postgraduate student should be capable of such a grotesque display of insubordination. 'In my opinion,' he sympathized, 'Dr. Schatz is behaving like an ungrateful spoiled, immature child. When he grows older, he will regret what he has done and is doing now.' William Feldman of the Mayo Clinic was among the few to express regret at Schatz's failure to win a share of the Nobel Prize. 'Your contribution' he wrote, 'was quite indivisible from the contribution of Doctor Waksman.' But even he counselled a stoical resignation to the committee's decision.

251

Persona non grata

In the years that followed, Schatz's successful legal battle seems to have made it easier for colleagues to efface him from the streptomycin story. Having made himself deeply unpopular, no one wanted to hear his point of view. Never afraid to take on his enemies, Schatz did whatever he could to fight back. But it was to no avail. Academic and popular journals lined up to reject his account of what actually happened. His testimony was eventually published in the *Pakistan Dental Journal* in 1965. Even then Schatz's side of the story emerged only because he was on the journal's editorial board.

Waksman later portrayed Schatz as a hardened ingrate. He was too self-obsessed, Waksman implied, to see that everything done in the Rutgers laboratory was a result not of the creativity of individual students but of Waksman's personal hands-on direction. In musical terms, Waksman saw himself as both composer and conductor: everyone played to his tune under his constant supervision. As a result, he interpreted every experimental success as his by proxy. 'Here I was about to be dragged into court,' Waksman fumed, 'at the instigation of one of my own students, whom I educated, [and] to whom I had pointed the way in the field of science.' Schatz, it seemed to Waksman, had bitten the hand that fed him. And this is the interpretation that nearly all Waksman's colleagues and more than a generation of medical historians endorsed. Schatz had spoiled the party by cynically cashing in on a better man's unworldly aversion to controversy.

The difficulty with this interpretation is that in 1950 the judge had not dismissed Schatz's claim out of hand. The settlement, which Waksman had accepted, made clear that Schatz had played an important role. But this was an objection that could be disposed of easily. A mere judge, it was argued, could hardly be expected to appreciate the subtleties of laboratory conventions. Luckily, a higher court of scientific justice existed, which could: the Nobel Prize Committee. Accordingly, they conferred this unparalleled scientific honour upon Waksman alone. As one historian put it a few years ago, 'Waksman felt he had been dealt with unjustly [in 1950] but his vindication came in 1952 when he became the sole recipient of the Nobel Prize.'

Whether this assessment is a fair one depends on the answer to a simple question: did the discovery of streptomycin require significant independent effort on the part of Albert Schatz?

No ordinary student

All the available evidence suggests that Schatz did indeed make a major individual contribution to the discovery of streptomycin. Paradoxically, our best evidence for this comes from the pen of Selman Waksman himself. As the biologist and historian, Milton Wainwright, has recently observed, in the weeks and months following the discovery there was no doubt in Waksman's mind that his graduate student deserved equal recognition with himself.

This is the only plausible way of explaining why Waksman allowed Schatz's name to appear before his own on the two 1944 papers that reported, first, the discovery of streptomycin, and then its efficacy against tuberculosis germs. When relations between the men later turned sour Waksman tried to wriggle free from this fact. Referring to the authorship of the 1944 articles, he coolly remarked that he was in the habit of giving students priority on significant papers so as to help advance their careers. As Wainwright notes, the difficulty with this is that the 1944 papers are the only occasions on which any student's name appeared before Waksman's on any paper, important or trivial.

There is also the evidence of the sworn affidavits. On 9 February 1945, Schatz and Waksman together signed legal documents confirming that they were the 'co-discoverers' of streptomycin. As the judge in *Schatz v. Waksman* recognized, it was very hard for Waksman and his attorneys to explain away the awkward fact that on two separate occasions the two men had sworn that, 'They verily believe themselves to be the original, first and joint inventors of an improvement in the same [i.e. streptomycin].' Fittingly in a year marking the return to peace, they had afterwards shaken hands as 'partners in streptomycin'. Nor was Waksman open-handed in the matter of patent applications. Elisabeth Bugie had been the second author on the first streptomycin paper, but having accepted that she hadn't played a critical role, she was required to sign an affidavit to this effect. In it Bugie unhesitatingly affirmed that both men deserved equal credit, referring to the compound that 'he [i.e. Waksman] and Dr. Schatz had discovered'.

Why did Waksman go to such lengths to acknowledge the contribution of his student? The answer is uncomplicated: at this stage he recognized that Schatz had been no ordinary doctoral student. First, Schatz had worked with an unusual degree of independence in the basement lab. It was only *after* the discovery of streptomycin that Waksman began to take a more active role in Schatz's work (and even this was limited because Schatz's experience meant that he needed little guidance). Second, it seems that, quite sensibly, Waksman had kept well clear of the lethal microbes with which Schatz was working. Third, Schatz had helped refine existing techniques in isolating the active factor produced by *Streptomyces griseus*. Last, having demonstrated the antibiotic properties of streptomycin, he had worked almost single-handedly to produce it in sufficient quantities for the first clinical trials to be undertaken.

No matter how covetous Waksman was of his status in the lab, he knew very well that Schatz had made a significant contribution. Of course, in carrying out his tests, Schatz stood upon the shoulders of many scientists of above average stature. He was also pursuing a research project initiated by Waksman, and using techniques developed by his boss. But Schatz worked at Rutgers with the independence and self-assurance of an established research scientist. Given this, Waksman's allowing Schatz the status of lead author on the first two papers seems to have been a simple matter of 'doing the decent thing'. It is his subsequent behaviour that needs explaining.

Waksman's wheeze

Things started to change when the full significance of the discovery dawned upon those involved. During 1946, as tuberculosis sufferers all over the world realized that they now had a lifeline, media interest in streptomycin intensified. A long series of magazines and newspapers ran dramatic stories about the development of the new miracle drug, the

'American penicillin'. Unintimidated by his new-found celebrity, Waksman accepted scores of invitations to speak to medical institutes and conferences throughout America, Europe, and later the Far East. As we've seen, in his last months at Rutgers Schatz was already beginning to feel that he was being sidelined by a more media-savvy boss. Press accounts increasingly ascribed the discovery to Waksman alone. Waksman himself sought to explain this to a disgruntled Schatz by claiming that the media had distorted his words. But a seed of doubt had been sown and, as we have seen, Albert Schatz departed from Rutgers a less than happy man.

Not long after, he wrote a letter to Waksman. Schatz's prose captures well how his formerly reverential feelings towards his old boss were gradually giving way to cynicism. He wrote:

> I feel particularly proud to have been associated with your group in the work on antibiotics, a subject which has raised the status of microbiology to a science second to none. In assisting you with the isolation of the streptomycin-producing organism and in the isolation of streptomycin itself, I feel that I have rendered my own contribution, no matter how small it may appear, to building and developing the science of antibiotics.

It was subtly done, but Schatz was staking his claim. It is highly unlikely that the sub-text was lost on Waksman.

Perhaps this is why Waksman sent Schatz a total of 1500 dollars over the following three years, all drawn from the streptomycin royalties he received. Each cheque for 500 dollars was accompanied by a request for Schatz to sign a letter waiving more rights to compensation for his part in the streptomycin discovery. Schatz had no objection to complying with these requests since he assumed that Waksman was making the same sacrifices. It slowly dawned on him that Waksman's real objective was to neutralize the affidavits of 1945. At best, the gifts of money were a salve for Waksman's conscience. More probably they were a sprat to catch a mackerel. Schatz's material gains were measured in hundreds of dollars, his losses in tens of thousands. It was on realizing this that Schatz wrote to Waksman.

Waksman, however, was unrepentant. In his reply of 28 January 1949 he conceded that Schatz had helped in 'the isolation of one of the two cultures of *S. griseus*' and assisted 'in the development of the methods for the isolation of the crude material from the media and in testing its antibacterial properties'. Even so, Waksman felt justified in launching a powerful counterattack, 'I hope you recognize,' he added, 'that this was only a very small part of the picture in the development of streptomycin as a whole.' The problem with this statement is that it is so blatantly contradicted by the listed authorship of the 1944 articles and the patent affidavits sworn the following year.

Subsequently, Waksman may have convinced himself that he'd agreed to share credit with Schatz in a brief moment of euphoric irrationality. But he must also have had some inkling that his ex-student had a reasonable claim for much greater credit. This is apparent from the manner in which he would soon begin tampering with the historical record. It is also suggested by the blatant threat with which his letter closed, 'You have made a good beginning as a promising scientist,' Waksman warned. 'You have a great future and you cannot afford to ruin it.' There is nothing here of the Duke of Wellington's admirable advice, 'Publish and be damned.' Waksman's meaning is clear: 'Make a fuss and I will destroy you.'

By the time he wrote again to Schatz, on 8 February 1949, Waksman had had more time to reflect. This had done nothing to improve his temper. With righteous indignation he raged:

> How dare you now present yourself as innocent of what
> transpired when you know full well that you had nothing to do
> with the practical development of streptomycin and were not
> entitled to special consideration . . . I must, therefore deny quite
> emphatically that I ever suggested or believed that you had any
> such rights.

Again, this statement is difficult to square with how important Waksman felt Schatz's work to have been in 1944 and 1945. It is also hard to believe that Waksman considered the 'practical development of streptomycin' the most crucial stage. Since most of the development work had been done by Feldman, Hinshaw, and the Merck scientists, sticking rigidly to this claim was tantamount to denying the importance of his own role in the streptomycin story.

There is an ad hoc, even clumsy, feel to Waksman's early attempts to shove Schatz off the podium. But, in the following years, he would tune his approach more finely. By the late 1960s it would be as if Schatz had never entered his lab.

The sick chicken routine

In his letters of January and February 1949 Waksman first aired a new claim on which his attorneys would place considerable weight the following year. It concerned the source of the *Streptomyces griseus* specimens from which Schatz isolated streptomycin. Schatz had always insisted that he had personally obtained both strains of *S. griseus*, one from manured soil and one from a fellow graduate student, Doris Jones. In January 1949, Waksman first begged to differ. Having never previously challenged Schatz's story, he now argued that he'd personally handed over to him one of the cultures, with clear directions for him to screen it against deadly tuberculosis germs. Waksman's intention here is obvious. He wanted to demonstrate that his graduate student was always following orders, that he scarcely blinked without first receiving instructions to do so from his boss. Over the next few months, Waksman conjured up a charming story to give his counter-claim the appearance of authenticity. This tale became a staple of newspaper reports of the discovery, giving the streptomycin drama a moment of rich fortuity to match that of Fleming's accidental discovery of penicillin.

It began with a sick chicken. Waksman's version started:

> One day early in August, 1943, a New Jersey farmer observed that one of his chickens seemed to be suffering from a peculiar ailment that affected its breathing. Fearing an epidemic of some kind, the farmer took the chicken immediately to the poultry pathologist at the nearby Agricultural Experiment Station.

By a circuitous route, Waksman wrote, an agar plate with several thriving populations of actinomycetes made its way to his laboratory. 'One of my graduate students was working at that time on the screening of actinomycetes,' he went on, '[so] I handed it to this student.' The unnamed student was obviously Albert Schatz. But lest anyone infer from this that Waksman had been merely a conduit, in a letter to the historian, S. Epstein, Waksman further embroidered his account. Having obtained the sick chicken's culture plate, he wrote, he took it to one of his assistants and issued a specific order. 'Schatz,' he wrote, 'was directed to transfer the three cultures, using standard procedure and [he] complied.' In light of what we now know, we can see that virtually every word in this passage was carefully selected to achieve the desired effect.

However, in the long term the effort failed. It's now clear that Waksman's sick chicken story was pure invention. We know this because, thanks to the labours of Milton Wainwright, we have the testimony of the student, Doris Jones (now Rolston), with whom the *S. griseus* sample to which Waksman referred actually originated. A good friend of Schatz's, Doris Jones was working in the laboratory of the Rutgers poultry pathologist, Frederick Beaudette, when she obtained a throat swab from a perfectly healthy chicken and isolated an interesting-looking actinomycete. Knowing that Schatz was screening these organisms for antibiotic effects, she walked across to his basement laboratory, tapped on the window, and handed him a Petri dish with a few distinctive colonies of actinomycetes. Jones confirms that neither Beaudette nor Waksman ever encountered the specimen until Schatz drew their attention to it. Jones' testimony has since been confirmed by the most unexpected of sources.

In 1989 Milton Wainwright unearthed a letter written by Waksman in the archives of Rutgers University. It was a response to a letter of May 1946 written by a Dr. R. A. Strong and requesting details of Albert Schatz's role in the genesis of streptomycin. Waksman seems to have felt at this stage that, in order to minimize Schatz's importance, it was enough simply for him to omit his name from the standard line he gave to the press. Then the sick chicken wasn't even a twinkle in an imaginary cockerel's eye.

As a result, in his letter to Dr Strong, Waksman simply described what really happened: how 'Assistant No 1' (Doris Jones) had 'brought the plate to assistant (No 2), Albert Schatz, without the involvement of the boss of either lab. It would be three years before the sick chicken and the anxious New Jersey farmer put in an appearance. By that stage the gloves were off.

But when *Schatz v. Waksman* came to court in 1950, it soon became clear to Waksman and his attorneys that this tale wasn't in itself enough to discredit Schatz's case. The problem was that the strain now being used by Merck to produce large quantities of streptomycin seemed to have absolutely nothing to do with Jones' chicken. According to Schatz, this second strain had been isolated from a sample of manured farm soil. Till now, Waksman had never challenged this assertion. And in his letter to Schatz of January 1949 he had acknowledged his role in 'the isolation of one of the two cultures of *S. griseus*'. Court proceedings were already underway when Waksman announced that Schatz was after all mistaken. In fact, he argued, the second specimen was no more than an offshoot from the sick chicken strain. This meant that, so long as the judge first bought his chicken story, Waksman could also claim to have been the source of the *S. griseus* sample used to produce streptomycin.

It's hard to believe that Waksman did not know the spuriousness of this claim. As Wainwright has argued, only a few months earlier Waksman himself had written in a scientific article that the lab's two *S. griseus* strains had always been kept far apart, 'excluding the possibility of one originating from the other as a contaminant'. In addition, the strains differed in nature. The second produced so much more streptomycin than the first that it's almost inconceivable that one could have budded from the other. To put it delicately, Waksman had changed his tune.

An anonymous assistant

When he flew to Oslo in 1952 to collect his Nobel Prize, Waksman was well aware of the campaign to have Schatz added to the list of recipients. And he arrived fearful lest the saga tarnish what ought to have been his proudest moment in science. Perhaps sensing the stress he was under, King Gustav VI is said to have whispered to Waksman, 'You have nothing to fear.' Even so, speaking before the great and the good at the award ceremony, Waksman played a cautious hand. Not wishing to leave any hostages to fortune, he skipped hurriedly over the circumstances of the discovery. Anyone in the audience hoping for a stirring account of sick chickens and a Eureka moment went away sadly disappointed, for

Waksman devoted less than thirty seconds to the events of September 1943. The few remarks he did make about these crucial days were couched in an anonymous, passive voice. '*S. griseus*,' he said, 'was first isolated in September 1943, and the first public announcement of the antibiotic was made in January 1944.'

Most of Waksman's speech concerned subsequent work on the clinical development and molecular analysis of streptomycin. Evidently, this was no triumphal oration. As he surely realized, the penalty of playing down Schatz was that he couldn't play himself up without risking further accusations. He did, however, mention Hinshaw and Feldman on several occasions. In contrast, Schatz's name appeared only once, in a list of twenty 'assistants and graduate students' appended to the end of the written version of his speech in a small acknowledgements section. This list could be taken to reflect how far Schatz had fallen in Waksman's estimation. The names are put down, one presumes, in order of relative importance. Schatz's name appears in twelfth position, behind several minor players and one (Elisabeth Bugie) who had explicitly denied being an important part of the process of discovery. From his status as first author on one of the most significant medical papers of the twentieth century, Albert Schatz had been demoted to a low-ranking name in a list that hardly anyone would read.

Professor A. Wallgren had the responsibility of introducing Waksman's discovery at the Nobel award ceremony. He did his best to avert further quarrelling by naming Schatz as the person who first witnessed the antibiotic effects of *S. griseus*. But Wallgren was far more committed to affirming the rightness of the Nobel Committee's decision than in redressing an alleged wrong. At every conceivable point Wallgren therefore drew attention to the significance of Waksman's prior work and to his hands-on supervision of his Rutgers graduate students. The audience heard that Waksman 'directed' Schatz's work, that he tested *S. griseus* for antibiotic effects alongside his student, and that at all times his students performed their work according to 'the clear principles which had been set out previously by Dr. Waksman'. Schatz became just another pair of hands.

But the image this conjured up of a wise supervisor standing at his student's side at the lab bench, commenting, encouraging, and advising, hardly fits what we know of the relationship between Waksman and Schatz at Rutgers. While Schatz unquestionably looked up to Waksman, his boss avoided the dangerous basement lab. Far from being responsible for testing streptomycin on tuberculosis germs, Waksman wasn't even in the vicinity when Schatz ran these vital but dangerous tests. But it was in both Waksman's and the Prize Committee's interests to sideline Schatz, and this is exactly what happened.

Yet for all Wallgren's deft verbal footwork, Waksman must have been pained by the Prize Committee's obvious sensitivity to claims that it had honoured the wrong man. One can imagine him wincing when Wallgren emphasized that the prize was being given not just for the discovery of streptomycin but also for Waksman's 'ingenious, systematic, and successful study of soil microbes'. Waksman surely realized that by emphasizing the research he'd done before 1943 and therefore without Schatz's help, the Nobel Committee was attempting to head off further controversy. Now the Committee, it seemed, considered its initial judgement questionable, if not definitely unsafe.

Downhill from Oslo

After Oslo, Waksman devoted more time to the rewriting of history. In 1958 he published a largely autobiographical book entitled *My Life with Microbes*. In this work, as in his Nobel speech, Waksman continued to move swiftly over the sequence of events that led to the discovery of streptomycin. Doing so allowed him to avoid any mention of the contribution made by Schatz without ever explicitly denying that he had played a key role.

Over the following few years, however, Waksman became increasingly bold. Schatz's campaign for recognition seemed to have foundered and he was now safely out of the way in a Chilean university. Now that the road seemed to be clear, Waksman brought out *The Conquest of Tuberculosis*, a book in which he finally provided a full account of September 1943. Far from making amends for past sins, Waksman perpetuated old distortions and added several more. He began by repeating the myth of the sick chicken, claiming that he'd received the plate of actinomycetes from Beaudette and 'handed it' to 'a student' to perform the appropriate tests. At all times Schatz was impersonally referred to as 'one of my graduate students'. More egregiously, Waksman now stated that once his 'student' had 'isolated and cultivated' the *S. griseus* specimens, he 'then asked another student to start several cultures ... so that an attempt could be made to concentrate the active material and to determine its nature'.

Behind this string of distortions lies the truth, as verified by Waksman himself in 1944 and 1945, that Albert Schatz, at considerable personal risk, alone isolated streptomycin, tested it against tuberculosis, and produced the first concentrated supplies used in clinical trials. But, determined to present himself as the one continuous thread running through the story, Waksman broke the discovery into several discrete steps and attributed each to a different student, all of whom were said to be dutifully following his orders. Waksman's description has the flavour of a victorious general telling how he commanded his army from a position atop a hill overlooking the battlefield. While his individual regimental commanders were wreathed in cannon smoke, and absorbed by immediate duty and present danger, he alone was able to observe the battle as a whole and issue orders accordingly. Only he, Waksman implied, had a clear grasp of the overall direction of the streptomycin research.

In *The Conquest of Tuberculosis* Waksman tendentiously confronted the difficult problem of assigning responsibility for a scientific discovery. Should one, he rhetorically asked, acclaim the chicken that 'picked up the culture from the soil' and developed a sore throat? Or perhaps the poultry pathologist who recognized that the swab from the chicken's throat contained an abundance of actinomycetes? Waksman's other contenders included, 'the [unnamed] student who worked under my direction and used the methods that I developed for the further cultivation and testing of these colonies'; 'my students and me in my laboratory'; 'my students working in the Merck laboratories'; and Feldman and Hinshaw for the clinical trials they undertook. These were carefully crafted sentences that hinted at a bigger truth, the real impossibility of confining the moment of discovery to a single individual's efforts. But they also deliberately concealed something much closer to home: the fact that without one particular student's dogged hard work and willingness to experiment with a lethal form of the tuberculosis bacteria, it might have been years before streptomycin was actually discovered.

With Schatz not in a position to press his case, Waksman's account achieved the status of revealed truth. On the fiftieth anniversary of the discovery in 1993, several newspapers and magazines repeated it almost verbatim. The *Smithsonian* went even further. It excised from the standard account any mention at all of Waksman's students and assistants. Swallowing whole the myth of the sick chicken, its correspondent described how Waksman personally 'analysed the barnyard soil and isolated the problem—a peculiar fungus. In the process,' the article continued, 'Waksman fortuitously discovered that the microorganism . . . produced a chemical that slowed the growth of certain bacteria.' Only in the last decade has the scale of Schatz's contribution started to be publicly recognized.

A feudal chieftain

The anthropological dimensions of Waksman's feud with Schatz have already been touched on in the introduction to this part of the book. To some extent at least, the clash arose because Waksman found it hard to adjust to the singular nature of his relationship with this particular underling. As a mature student who didn't need close supervision and was working alone in an indisputably dangerous laboratory, Schatz had much more freedom than Waksman's other assistants. When he made his crucial discovery, Schatz knew that he was entitled to a share of the recognition. Waksman, however, found it increasingly difficult to confer on him any more credit than he would ordinarily grant a research student. That this is a problem extending well beyond the relationship between these two men is reflected in Wainwright's finding that it was mainly junior scientists who supported Schatz and, disproportionately, senior personnel who backed Waksman.

The seigniorial outlook of a head of department cannot, however, provide a full explanation for the enmity Waksman felt towards Albert Schatz. After all, in both 1944 and 1945 the senior man did show sufficient mental suppleness to recognize his junior's independent efforts. It's therefore hard to avoid the conclusion that Waksman sidelined his student so that he himself could monopolize the kudos once it became clear that streptomycin (used in conjunction with another compound called PAS) might well vanquish tuberculosis in the Western world. But there's also a difficulty with this straightforward, if cynical, reading: Waksman always ensured that when he put his own name to the fore those of Feldman and Hinshaw accompanied it. He even expressed the opinion that his Nobel Prize ought to have been shared with them. This accords with most people's recollections of Waksman as a very decent person. Why then did he feel the need so ruthlessly to write Schatz out of history?

Reputation and reality

Sometime in 1945 Waksman realized that streptomycin royalties might fund what he most coveted, his own institute at Rutgers. To avoid having to split his part of the profits, he agreed to share the credit with his student but pretended to be forgoing his entitlement to any income from the discovery. Without Schatz's knowledge, he then started amassing large sums in royalties that he earmarked for the construction of what became the Rutgers Institute of Microbiology, later renamed in his honour. Not wanting anyone to pose embarrassing questions about why Schatz wasn't receiving royalties, he also began de-emphasizing Schatz's role in the discovery. If Waksman had qualms, the worthiness of the cause no doubt helped him stifle them. But financial gain was not the only consideration.

Waksman was surely aware that if Schatz's role became well known, he risked being relegated to a secondary status. After all, the Schatz story made far better press copy than his own. Here was a young man who gave up his free time in the military to trawl swamps for bacteria, who locked himself away in a lab with a highly dangerous pathogen to fight a killer disease, and who did all that could be asked of him and more to bring the new-found miracle drug through the test phase. He would have fitted the bill for a classic, all-action American hero to a greater extent than his not overly colourful boss.

Against this backdrop, Waksman received the unexpected letter from

Albert Schatz. At this stage, as he may later have come to see, he ought to have put the matter right. But by then the stakes were very high. Perhaps it was this realization that drove Waksman to risk allowing the matter to come before a public court. Either way, it was a dreadful miscalculation. The verdict, as Waksman later wrote, turned 1950 into the 'darkest' year of his 'whole life'. With his good name compromised, he now had more need than ever to efface Schatz from the historical record, both to bury a controversy that had marred his greatest scientific triumph, and to prove to himself and to others that the trial verdict had done him a gross injustice. Having boxed himself into a corner, this ordinarily decent and generous man now behaved unusually badly.

But for Waksman there may have been another difficulty with Schatz's involvement. As we've seen, iconic discoveries in science are nearly always associated with one or more of the following: an act of astonishing intellectual creativity, a brilliantly conceived experiment, or a chance event with revolutionary consequences. Streptomycin was undoubtedly a tremendous boon for humanity, but its discovery conformed to none of these criteria. In setting up his screening programme, Waksman had no need for creative genius. Others had already demonstrated that harmless micro-organisms can kill lethal bacteria inside the human body. Nor were his experimental methods radically innovative. For the trained biochemist, looking for antibiotics in Waksman's lab was not unlike panhandling for gold, requiring far more patience than sustained intellectual effort. And, in spite of his spurious sick chicken story, serendipity was never a significant factor. The key micro-organism was out there in abundance and its identification was just a matter of time, dedication, and courage.

In 1945 the press proceeded to build Waksman up into a kind of Albert Einstein of biology. Knowing what they were after, Waksman manufactured a fittingly heroic drama. He exaggerated the number of soil microbes it had been necessary to test before they found streptomycin (giving his research, if nothing else, the appearance of Herculean labour). And on various occasions he enigmatically alluded to his having introduced 'a new philosophy' that played a key role in his team's success. Yet therein lay another problem for Waksman. Who would believe that the discovery had required a bold intellectual thrust into the unknown once they found out that a mere graduate student had played a lead role in it? Of course, Einstein had been a patent clerk when he first published his ideas on relativity. But so manifestly revolutionary were his theories that this only added drama to his story. The brilliance of Waksman's discovery didn't speak for itself in the same vivid way. At the very least, then, all scenes involving Albert Schatz had to be dropped from the canonical script Waksman wrote.

Although rewards in science go to towering geniuses, lucky plodders, and anything in between, this is often not what we want to be told. Each great breakthrough requires a hero. Responding to this demand, Waksman strove to live up to a romantic perception of the nature of scientific discovery that in his case simply didn't fit the facts. Perhaps, in part, Albert Schatz was sacrificed to this dubious end. Yet Waksman's screening programme for soil antibiotics ought to have been held up as an object lesson in efficient and well-conceived biological research. His goal was of vital importance and his method optimally geared to attaining it. Even if his endeavour did lack the glamour of theoretical physics, it had something at least as important to recommend it: streptomycin helped to save hundreds of thousands of lives.

THE BIGGER PICTURE

Recognizing that scientific discovery is a community process is an important first step towards proper contextualization. But dividing up the spoils more fairly is only a start. Having dismantled the myths, it's also vital for the historian to examine how particular social contexts have shaped the formulation of scientific paradigms. This endeavour has acquired a rather bad press in recent years. It has also generated considerable enmity between humanities and science faculties on many university campuses. Many people seem to feel that relating scientific theories to anything beyond the laboratory door is to consider scientific knowledge no more reliable than theology, poetry, or political polemics. I think this an overly defensive position to take, for, as I hope to show in these closing lines, to say that a scientific theory is linked in some way to a particular social context is to imply nothing, one way or the other, about its likely truth value.

More to the point, to deny the importance of social context is to close off the possibility of understanding how the modern scientific enterprise itself came into being. Science may seem like applied common sense ('common sense at its best' is how the Victorian biologist, T. H. Huxley, described it), but for centuries it was anything but obvious that one should search nature for new truths or subject rival theories to experimental testing. During the Renaissance, in fact, most of Europe's natural philosophers believed that pure knowledge was to be derived not from studying nature directly but by poring over ancient and classical texts in search of lost or hidden knowledge.

It was only with the growth of mercantile and then industrial capitalism that modern science became feasible. The connection between capitalism and science remains an area of intense debate among scholars, but most agree that with the demise of feudalism there appeared new social structures, forms of production, and cultural

267

attitudes more conducive to the effective study of nature than anything that had gone before. Among the changes arguably facilitating the emergence of modern science were the production of surplus wealth that could be used to support scientific research; the development of industries demanding and rewarding the exploration of physical, chemical, and biological phenomena; the tremendous expansion in our knowledge of the world that came with the growth in global trade and exploration and which led European scholars to question the credibility of ancient and classical authors; and the development of more and more effective communications permitting the propagation of ideas and methods. These conditions promoted and were in turn promoted by the appearance of a utility-minded class of men, and later women, committed to improving life on earth by actively changing it rather than passively awaiting the joys of a heavenly paradise. Such factors, combined with a series of technological advances, made possible the extraordinary range of discoveries that define the Scientific Revolution.

Glanvill, Newton, Lind, and Spallanzani were all direct beneficiaries of the dramatic social, economic, and ideological shifts experienced by Europe in the 1600s and 1700s. But their world-views were also more directly conditioned by changing patterns of production and subsistence. Already by the early 1600s, improvements in agriculture, the large-scale felling of pristine forests, the growth of towns and cities, and the development of industrial technologies had created micro-environments in which nature no longer appeared so capricious. More than ever before it seemed that natural forces could be understood, exploited, and controlled. At the same time, the introduction of proper book-keeping methods, the recording of births and deaths, and the growth in other forms of bureaucracy revealed more consistent patterns beneath what had once seemed haphazard natural phenomena. The emergent feeling that nature was intelligible, regular, and predictable in turn encouraged the view that the universe was governed by fixed, natural laws. Descartes' metaphor of the clockwork universe exemplifies the intimacy of the relationship between natural philosophy and technological advancement. And, inspired in much the same way as Descartes, many other natural philosophers of the period searched for basic laws of nature. This endeavour paid handsome dividends in, among other achievements, the discoveries of Newton, Boyle, and Hooke.

Once nature was seen as an embodiment of divine order, any attempt to explain natural events in terms of mysterious or occult forces was liable to be branded as superstitious. It was in this context that Spallanzani refuted epigenesis as superstitious nonsense and embraced ovism as the only viable alternative. Lind's view of the body as a sophisticated machine prone to faulty plumbing was another application of the machine metaphor.

No less important, the increasing mastery over nature achieved from the early 1600s helps to explain the tremendous self-confidence of natural philosophers, the irrepressible self-belief that led them to think they could explain all nature's mysteries, from magnetism and generation to planetary motion and witchcraft. Not all Royal Society philosophers were as convinced as Joseph Glanvill that witches were real, but his conviction that the Devil had special powers, that these were essentially natural, and that they were susceptible to scientific explication must be seen as one of the clearest expressions of the self-belief that marked out the new generation of natural philosophers.

Historical circumstances also shaped the scientific revolution and its aftermath in more specific ways. For all of Newton's machinations at the Royal Society, the acceptance of his theory of light and colour critically depended upon eighteenth-century improvements in the standards of glass manufacture. But these advances were not made at the behest of natural philosophy: science benefited only incidentally from the progressive refinement of all manner of manufacturing processes. The Royal Society's adoption of the experimental method was also in part determined by wider developments.

As noted in an earlier chapter, having just emerged from years of bloody internecine conflict, the English social elites of the 1650s had little appetite for further conflict. So they devised a method for doing science, based on experiment and collective witnessing, that aimed to maximize consensus and reduce the probability of debate degenerating into ugly feuding. In this context, the experimental method acquired tremendous appeal.

It is also worth pointing out that the Civil War was itself related to realignments in English society generated by, among other things, a mercantile expansion dating back to the Tudors. The same economic trend was yet more important to the development of Lind's views on the causes and cure of scurvy. Britain experienced a massive surge in foreign trade during the 1700s, requiring the setting up of a vast infrastructure of warships and personnel, not to mention dozens of far-flung ports, strategically placed garrisons, seaside hospitals, and foreign consulates. As Britain's wealth and security came to depend more and more on seaborne trade, the highest of premiums was set upon finding reliable prophylaxes and cures for this plague of the sea. This is why Lind first turned to the study of the disease and why he was so handsomely rewarded after the publication of his 1753 treatise. Needless to say, warfare added further impetus to the search for effective anti-scorbutics. It was, of course, the Austrian War of Succession that gave Lind his first opportunity to experiment with citrus fruits, and international conflict created the conditions in which the true value of citrus fruit as an anti-scorbutic was properly recognized during the 1790s.

By the 1800s scientific advance remained closely linked to political and economic developments. As the efficient harnessing of power became of vital economic importance with the onset of industrialization, physicists turned from the study of matter to that of energy, resulting in brilliant new understandings of the equivalence of forces and the properties of energetic systems. The period's emphasis on quantification in science was also linked to the growing importance of fine-scale calibration in manufacturing.

Medicine too changed in response to industrialism. While Pettenkofer's fame lasted, it was critically dependent on two things: first, the growth of large and insanitary towns and cities; and second, the fact that many European states during the 1800s had begun to accept responsibility for the physical well-being of their citizens. Having calculated that national wealth and security required a healthy workforce and physically robust army recruits, states began to encourage studies of the environmental determinants of epidemic disease. Several then embarked on a long process of sanitary reform.

It's ironic that Pettenkofer's popularity suffered from the increasing interventionism that followed the unification of Germany in 1871. But here too political imperatives obtruded into scientific debate. Bismarck and the Emperor Wilhelm I sidelined Pettenkofer and elevated Robert Koch partly because the latter's germ theory of disease emphasized the contagiousness of epidemic disease and thereby justified the imperial authorities interfering in the affairs of individual states. As a result, Koch's ideas could win through years before there was sufficient empirical evidence to invalidate Pettenkofer's rival theory. With respect to Koch's scientific achievements, it should be noted that bacteriology could have gone nowhere without the availability of new dyes, produced by textile companies, which happened to be able to stain both microbes and garments.

The fate of Semmelweis, like that of Pettenkofer, was also sealed by a reassertion of imperial authority. This tragic Hungarian was destroyed by Viennese hospital authorities after he'd taken an active part in the failed revolution of 1848. His boss, Johannes Klein, would not suffer having a committed democrat, especially a Hungarian democrat, working in his wards. For Klein, Semmelweis and everything he stood for became intolerable; therefore he had to go. Flaws in Semmelweis' theory and data ensured that his hand-washing practice was not widely adopted elsewhere, and he achieved heroic status only because the British surgical profession of the late 1800s was alert for opportunities to improve its public image. Semmelweis, leading surgeons recognized, could easily be made to fit the template of the true intellectual hero.

Psychiatrists were even more creative than surgeons in manufacturing their heroes. Their need, after all, was the greater. Certainly, late Victorian surgeons struggled to shake off the stigma of having been joined in trade with barbers just a century or so before, and many people still associated their art with unbelievable agony, filthy instruments, and blood-stained sawdust. Yet by the late Victorian age surgeons were also being celebrated for the tremendous advances they had made in reducing levels of pain and infection, and for vastly expanding the repertoire of surgical procedures. In contrast, the professional status of psychiatry during the 1800s was precarious.

Creating a succession of romantic heroes in a struggle to make madness the province of medicine was one of the strategies they adopted. Another involved developing a set of theories that (as astute

271

contemporaries observed) had little empirical credibility but were highly effective in challenging the authority of lay and clerical asylum custodians. It is not unreasonable to say that the establishment of psychiatry as a medical specialty owed much to the ability of its practitioners to delude themselves into thinking that treatments like bleeding and purging actually worked and that mental illness was usually the result of cerebral injuries and swellings. And it was the same professional aspirations that disposed many practitioners to embrace phrenology, a framework of ideas that would otherwise have been rejected as scientifically dubious and, according to all but the most freethinking, theologically odious.

For much of the twentieth century psychiatry continued grappling for a paradigm both to unite its practitioners and to persuade the public that it had special expertise in the treatment of the mentally ill. By the mid-century, however, most other branches of medicine had entered a golden age. Thanks to large-scale investment by government and industry, medical science had finally come of age. Research centres were now attached to many American and European universities and hospitals. And, with the advent of sulfa drugs, a variety of vaccines, and above all penicillin, these centres had borne wonderful fruit. It would have taken astonishingly good luck for the antibiotic properties of streptomycin to have been discovered during the nineteenth century, before the shift in the locus of medical inquiry from the ward to the laboratory. But Waksman and Schatz worked in a lab generously endowed (in comparison to earlier decades) with equipment and trained staff, with the result that streptomycin was almost bound to be isolated once attention had been turned to the medical potential of soil micro-organisms. The coming of age of medical science made the discovery not only possible but probable.

At the same time, the expansion of the pharmaceutical industry, dominated by firms like Wellcome, Merck, and Bayer, permitted streptomycin's rapid synthesis and development. War too played an indirect but important role in their willingness to help. That funding from private industry became available during the 1940s was in no small part due to the fear that the end of World War II might witness the same catastrophic epidemics as had killed millions after the Great War.

Global conflict thus hastened the introduction of streptomycin, just as the Napoleonic Wars had led to the introduction of citrus fruits to English sailors. If warfare isn't always the mother of invention, it has frequently acted as the midwife to significant scientific and medical developments.

This obviously applies to the genesis of Robert Watson-Watt's CH radar system. The threat of imminent German bombing raids galvanized the British government during the mid-1930s into converting a tool of meteorological inquiry into a technology without which Britain might well have fallen to the Nazis in the early 1940s. Hülsmeyer had the misfortune to introduce his radar contraption in 1905, long before anyone could predict its value. In contrast, Watson-Watt's famous memorandum was delivered to a government desperate to find a means of detecting incoming aircraft, and willing to put its all into any project with fair promise. Radar's extraordinary effectiveness in both attack and defence was sufficient to prove that ships and planes, whether civilian or military, could no longer afford to do without it. By the mid-1930s radar was here to stay.

In this brief survey it has been possible to identity several ways in which social factors, broadly defined, made an impact upon the scientific ideas looked at in this book. We've seen that particular combinations of social, economic, political, and ideological circum-stances can hasten or retard scientific progress. The wider context can also influence the sort of theories that arise, the kinds of metaphors available, the avenues of inquiry that seem most inviting, and the likelihood of specific theories gaining popularity. And, as the fates of Semmelweis, Pettenkofer, Weyer, and Pinel illustrate, the perceived interests of governments and professions sometimes play central roles in determining whether an individual becomes and then remains a hero of science. Simply put, the story of science cannot be told as if the world outside the laboratory did not exist. It's about much more than microscopes, test-tubes, and the application of pure reason (however that might be defined).

Especially in the human sciences, external factors can also help to determine the content and not just the status of theories. This certainly applies to early nineteenth-century psychiatry's fondness for phrenology and *anatomie pathologique*. Numerous others cases could be cited. In fact, virtually any discipline concerned with human mind and behaviour entails built-in, unproven, and often unexamined assumptions about the plasticity of human nature. Most such disciplines have recent histories in which the dominant theories were infused with sectarian racial or sexual prejudices; many continue to be subject to faddism and the swinging pendulum of opinion. But to repeat a point made earlier, just because a discipline is susceptible to outside influences does not mean that it's unfit for scientific investigation. Take, for example, Charles Darwin's evolutionary theory.

It's probably no coincidence that Darwin came up with the idea of natural selection at a time when England was undergoing a period of rapid technological innovation in which many people faced an increasingly harsh struggle for jobs, markets, and bread. Familiar with the grim realities of urban existence and the ubiquity of struggle from reading the works of Adam Smith and Thomas Malthus, and through observing the changing world around him, Darwin was able to perceive that nature too is a battleground in which no quarter is given. Eighteenth-century naturalists, living in comparatively gentler times, had seen harmony and God's infinite generosity where Darwin saw death to all but the best. Industrialism was thus a stimulus that made possible one of the key insights of modern biology.

Failing to take into account the wider context is to convey a false idea of science as involving the sterile accumulation of theories and facts. It's to suggest that science is somehow hermetically sealed off from the rest of society. Of course, one can attach too much importance to the wider world. Some scholars have tried to implicate cultural trends in the development of quantum theory and core aspects of both embryology and immunology. Such attempts seem to me far-fetched. We also need to remember that for all the importance of context, well-attested scientific theories can often be transferred from one cultural milieu to another with little if anything being lost in the translation (there are, though, striking cases where local scientific traditions have skewed the interpretation of imported theories). Moreover, although political factors might accelerate the adoption of an idea, the ultimate arbiter of whether an idea survives or not is the judgement of scientific researchers engaged in more and more rigorous experimentation.

Thus even if the acceptance of Robert Koch's germ theory was assisted by the politicking of Bismarck and Wilhelm I, we need to acknowledge that Pettenkofer's rival germ-soil hypothesis was always destined to be abandoned because it was wrong, and sooner or later would be proven so. While all but the most basic assertions in science remain provisional, at their best they're vastly more reliable than claims arrived at in less systematic ways. However, this does not alter the fact that certain ideas have been much more likely to arise and to strike a sympathetic chord in certain times and places than in others. Social, economic, and political factors introduce a degree of contingency that must be factored into our accounts of scientific discovery.

In this book I've attempted to negotiate a path between the extremes of relativism and a naïve reverence for the textbook version of the scientific method. I believe that we should make every effort to find common ground between these two extremes. The 1990s saw a bruising encounter in the world of academia between arts and science faculties, which was dubbed at the time, rather too grandiosely, the 'science wars'. Regrettably, the debate became polarized between a minority faction of relativists and a nucleus of scientists who tended to exaggerate the danger these individuals posed to the integrity of science. The camps also adopted unhelpfully extreme positions. Many of the 'social constructivists' who were involved expressed considerable cynicism about the explanatory power of science. But paradoxically, even bizarrely, they felt boundless confidence in the cogency of theories drawn from the social sciences. For their part, many of the scientific combatants fell back on claims about the purity of science that even 'realist' philosophers of science abandoned years ago.

Now that the dust has settled, it's to be hoped that both sides can seek a more nuanced view of the relationship between science and society, one that accepts the subjective components of science, the fragility of scientific knowledge, and the destructive as well as constructive aspects of its grounding in particular social environments. At the same time, we need constantly to keep in mind that for all the difficulties involved in doing good science, the experimental method has an unrivalled capacity to increase our understanding of the world around us.

CHAPTER 1: JOSEPH GLANVILL: SCIENTIFIC WITCH-FINDER

A few short books and articles have been written on Joseph Glanvill, all of which to some degree cover his interest in witchcraft: R. M. Burns' *The Great Debate on Miracles: From Joseph Glanvill to David Hume* (Lewisburg, Penna., 1981), Thomas Harmon Jobe's 'The Devil in Restoration Science: The Glanvill–Webster Witchcraft Debate', in the journal *Isis* 72 (1981), 343–56, Moody E. Prior's, 'Joseph Glanvill, Witchcraft, and Seventeenth Century Science', in *Modern Philology* 30 (1932), 167–93, and Sascha Talmor's *Glanvill: The Uses and Abuses of Scepticism* (Oxford, 1981). Brian Easlea's *Witch Hunting, Magic and the New Philosophy: An Introduction to Debates of the Scientific Revolution 1450–1750* (Brighton, 1980) is also valuable. Reprints of Glanvill's main works are also available, including *The Vanity of Dogmatizing*, with a critical introduction by Stephen Medcalf (Brighton, 1970) and *Saducismus Triumphatus or, Full and Plain Evidence concerning Witches and Apparitions* (Gainesville, Fla., 1966, facsimile edition). Very useful for getting a handle on the status of the idea of immaterial forces in seventeenth-century thought are Keith Hutchison's 'Supernaturalism and the Mechanical Philosophy' in the journal *History of Science* 21 (1983), 297–333, and his 'What Happened to Occult Qualities in the Scientific Revolution?' in *Isis* 73 (1982), 233–53. Also important in this regard are: John Henry's 'Occult Qualities and the Experimental Philosophy: Active Principles in Pre-Newtonian Matter Theory' in *History of Science* 24 (1986), 335–81, and Stuart Clark's *Thinking with Demons: The Idea of Witchcraft in Early Modern Europe* (Oxford, 1997).

CHAPTER 2: THE MAN WHO MADE UNDERPANTS FOR FROGS

For a modern take on the history of preformationism see: Clara Pinto-Correia's first-rate *The Ovary of Eve: Egg and Sperm and Preformation* (Chicago, 1997), John Farley's *Gametes and Spores: Ideas about Sexual Reproduction, 1750–1914* (Baltimore, Md., 1992), Shirley A. Roe's *Matter, Life, and Generation: Eighteenth-Century Embryology and the Haller–Wolff Debate* (Cambridge, 1991), and the collected papers in the volume *Lazzaro Spallanzani e la Biologia del Settecento* (Florence, 1982). Also useful is Elizabeth Gasking's *Investigations into Generation 1651–1828* (London, 1967). There is an interesting discussion of preformationism in Lewis Wolpert's *The Unnatural Nature of Science* (Cambridge, Mass., 1998), though to my mind his criteria for distinguishing science from non-science are implicitly presentist. See also: Iris Sandler, 'The Re-examination of Spallanzani's Interpretation of the Role of the Spermatic Animacules in Fertilisation', *Journal of the History of Biology* 6 (1977), 193–223.

CHAPTER 3: PETTENKOFER'S POISONED CHALICE

There are very few English-language studies of Max von Pettenkofer. The only book devoted to him is Edgar Erskine Hume's *Max von Pettenkofer: His Theory of the Etiology of Cholera, Typhoid Fever and other Intestinal Diseases: A Review of his Arguments and Evidence* (New York, 1927). There is also a translation of two of his lectures, together with a useful introduction, by the historian Henry E. Sigerist, published under the title of *The Value of Health to a City: Two Lectures Delivered in 1873* (Baltimore, Md., 1941). For a general account of nineteenth-century public health, the role of Pettenkofer, and his eclipse by Koch, see Richard J. Evans' *Death in Hamburg: Society and Politics in the Cholera Years, 1830–1910* (Harmondsworth, 1991), Dorothy Porter's *Health, Civilization, and the State: A History of Public Health from Ancient To Modern Times* (London, 1999) and George Rosen's *A History of Public Health* (Baltimore, Md., 1993 edition). In writing this chapter I have benefited from communications with Sandy Cairncross and Menno Bouma of the London School of Hygiene and Tropical Medicine.

CHAPTER 4: SIR ISAAC NEWTON AND THE MEANING OF LIGHT

This chapter is based on an analysis of Newton's theory of light and its reception conducted by the Cambridge University historian of science, Simon Schaffer, and published in his paper 'Glass Works: Newton's Prisms and the Use of Experiment', in David Gooding, Trevor Pinch and Simon Schaffer (eds), *The Use of Experiment: Studies in the Natural Sciences* (Cambridge, 1989), pp. 67–104. For a general account of Newton's life, work, and personality see Richard S. Westfall's *Never at Rest: A Biography of Isaac Newton* (Cambridge, 1980).

Patricia Fara's *Newton: The Making of a Genius* (London, 2002) explores his 'canonization' and representation as a scientific genius. The theme of experimenter's regress is discussed at length in Harry Collins and Trevor Pinch's *The Golem: What you Should Know about Science* (Cambridge, 1998). For further information on Newton's theories of light and colour see the articles by Alan Shapiro (e.g. 'Artists' Colors and Newton's Colors', *Isis* 85 (1994), 600–30).

CHAPTER 5: DR JAMES LIND AND THE NAVY'S SCOURGE

Kenneth J. Carpenter's *The History of Scurvy and Vitamin C* (Cambridge, 1998) is the best general history of attempts to explain and treat scurvy. A detailed account of Lind's work is provided in Michael Bartholomew's fine article 'James Lind and Scurvy: a Re-evaluation', published in the online *Journal for Maritime Research* (January 2001). For the popularity of rival remedies see William M. McBride's article, ' "Normal" Medical Science and British Treatment of the Sea Scurvy, 1753–75', *Journal of the History of Medicine and Allied Sciences* 46 (1991), 158–77. For an analysis of the way in which naval doctors continued to see fruit as only one aspect of treating scurvy into the later 1700s, and the role of discipline and hierarchy in naval attitudes to shipboard maladies, see Christopher Lawrence's 'Scurvy, Lemon Juice and Naval Discipline 1750–1815', in *The Impact of the Past upon the Present: Proceedings of the Second National Medical History Conference, Perth, July 1–5 1991*, ed. Peter M. Winterton and Desmond L. Gurry (Perth, Western Australia, 1992), pp. 227–32.

CHAPTER 6: THE DESTRUCTION OF IGNAZ SEMMELWEIS

Each of the following works sheds important new light on the life and death of this tragic figure: Irvine Loudon, *The Tragedy of Childbed Fever* (Oxford, 2002), K. Codell Carter and Barbara R. Carter, *Childbed Fever: A Scientific Biography of Ignaz Semmelweis* (Westport, Conn., 1994), K. Codell Carter, Scott Abbott, and James L. Siebach, 'Five Documents relating to the Final Illness and Death of Ignaz Semmelweis', *Bulletin of the History of Medicine* 69 (1995), 255–70, and György Gortvay and Imre Zoltán, *Semmelweis: His Life and Work* (Budapest, 1968). See also the chapter on Semmelweis in Sherwin B. Nuland's *Doctors:*

The Biography of Medicine (New York, 1998). At the time of writing a new book on Semmelweis has been published which also goes beyond the myth: Sherwin B. Nuland, *The Doctors' Plague: Germs, Childbed Fever, and the Strange Story of Ignac Semmelweis* (New York, 2003).

CHAPTER 7: WILL THE REAL JOHANN WEYER PLEASE STAND UP?

For more on Johann Weyer consult the introduction to Benjamin G. Kohl and H. C. Erik Midelfort's edited book *On Witchcraft: An Abridged Translation of Johann Weyer's* De Praestigiis Daemonum (Asheville, NC, 1998, translated by John Shea) and Patrick Vandermeersch's essay, 'The Victory of Psychiatry over Demonology: The Origin of the Nineteenth-Century Myth', *History of Psychiatry* 2 (1991), 351–63. For witchcraft more generally see Sydney Anglo (ed.), *The Damned Art: Essays in the Literature of Witchcraft* (London, 1977), Robin Briggs, *Witches and Neighbours: The Social and Cultural Context of European Witchcraft* (New York, 1996), and Stuart Clark, *Thinking with Demons: The Idea of Witchcraft in Early Modern Europe* (Oxford, 1996). In addition, the following article sheds much light on how our forebears conceptualized mental illness: Richard Neugebauer, 'Medieval and Early Modern Theories of Mental Illness', *Archives of General Psychiatry* 36 (1979), 477–83. See also: Thomas J. Schoeneman, 'The Mentally Ill Witch in Textbooks of Abnormal Psychology', *Professional Psychology: Research and Practice*, 1984, 15, 229–314.

CHAPTER 8: PHILIPPE PINEL: the reforging of a chain-breaker

For further information on Philippe Pinel and the chain-breaking myth see Gladys Swain, *Le Sujet de la folie: Naissance de la psychiatrie* (Paris, 1997), Doris B. Weiner, ' "Le Geste de Pinel": The History of a Psychiatric Myth', in *Discovering the History of Psychiatry* (New York, 1994), and Jan Goldstein's *Console and Classify: The French Psychiatric Profession in the Nineteenth Century* (Chicago, 1991). For an account of British psychiatry in the same period see the book edited by Andrew Scull, Charlotte MacKenzie, and Nicholas Hervey, *Masters of Bedlam: The Transformation of the Mad-Doctoring Trade* (Princeton, NJ, 1996). Roy Porter's *Madness: A Brief History* (Oxford, 2002) provides a general introduction to changing ideas about mental ill-health, as does Edward Shorter's *A History of Psychiatry: From the Era of the Asylum to the Age of Prozac* (new York, 1997). For more on the appeal of phrenology to radicals and the urban middle classes of nineteenth-century Britain see Roger Cooter, *The Cultural Meaning of Popular Science: Phrenology and the Organization of Consent in Nineteenth-century Britain* (Cambridge, 1985).

CHAPTER 9: THE FIRST CASUALTY OF WAR

This chapter is based on David Prichard's *The Radar War: Germany's Pioneering Achievement 1904–45* (Wellingborough, 1989), manuscript holdings relating to Arnold Wilkins held at Churchill College, Cambridge, and Robert Buderi's *The Invention that Changed the World: How a Small Group of Radar Pioneers Won the Second World War and Launched a Technological Revolution* (New York, 1996). See also the work compiled by Colin Latham and Anne Stobbs entitled *Pioneers of Radar* (Stroud, 1999).

CHAPTER 10: RANK HATH ITS PRIVILEGES

The chief source for this chapter is Milton Wainwright's article 'Streptomycin: Discovery and Resultant Controversy', published in *History and Philosophy of the Life Sciences* 13 (1991): 97–124. For general histories of tuberculosis and antibiotic research, including the streptomycin story, see Wainwright's *Miracle Cure: the Story of Penicillin and the Golden Age of Antibiotics* (Cambridge, Mass., 1990), Thomas Dormandy's *The White Death: A History of Tuberculosis* (London, 1999), and Frank Ryan's *Tuberculosis, The Greatest Story Never Told: the Human Story of the Search for the Cure for Tuberculosis and the New Global Threat* (Bromsgrove, 1992).

A

Academic Legion, the 146
Ackerknecht, Erwin 208
actinomycetes 246, 247, 248, 254,
    256, 257, 258, 259, 260, 262
Acton, Lord (John E. E. Dalberg) 219
Admiralty, the 114, 115, 117–121,
    130
*Aetiology, Concept and Prophylaxis of
    Childbed Fever* (Semmelweis)
    141, 153, 155
Agrippa, Heinrich Cornelius 180,
    182
An Essay on Fevers (Huxham) 118
Anaximander 10
Anson, Lord George 103, 118, 121
Apollo 13 165
Apollo, statue of 200
Appleton, Edward Victor 226, 227,
    230–234, 236, 238–240
Archimedes 85
Aristotelian philosophy 5, 11, 54, 55,
    60
Aristotle 54
Arneth, Franz Hector 143–144
artificial insemination 44–45
Ascent of Man, The (Bronowski) 9
Aura seminalism 46
Axenfeld, Theodor 187

B

Bacon, Francis 10, 15, 48
Bainbridge-Bell, L. H. 237

Baldwin, Stanley 220
Bartholomew, Michael 115
Bastille Day 192, 193
Battle of Britain 219, 220, 235,
    239
Bayle, A. L. J. 209, 210
Beaudette, Frederick 258
Bell, Jocelyn 216
Bennett, Sir Thomas 26
Bethlehem Hospital (Bedlam) 196,
    211
Bicêtre Hospital (France) 192, 193,
    195, 197, 198, 199, 201
Bismarck, Otto von 70, 77, 78, 80,
    271, 255
Black, Joseph 123
Blane, Sir Gilbert 114, 125, 130
Bonaparte, Emperor Napoleon 77,
    142, 192, 207
Bonnet, Charles 47, 51, 52, 55, 56,
    58
Boorstin, Daniel 165
Bourbons, the 195, 207
Bourneville, Désiré 187, 188
Bowen, Edward G. 232, 234, 236,
    237
Boyle, Robert 15, 16, 17, 22, 23, 28,
    29, 30, 31, 35, 36,174, 269
    and Joseph Glanvill 15–17, 24
    research on properties of gases
        23–24, 28, 35
    existence of occult forces 30, 31
Braun, Karl 148, 150, 151, 154, 155,
    158
Braun, Karl F. 223
Braveheart 165

Breit, Gregory 228, 232
British Broadcasting Corporation
	(BBC) 226
Bronowski, Jacob 9
Brooks, Jane 21
Brosier, Marthe 18, 25
Broutet, Monsieur 205
Browne, W. A. F. 210
Bruneval (France) 229
Bugie, Elizabeth 248, 253, 260
Burke, Edmund 201
Bury, J. B. 15, 17
Byron, John 118, 119–120, 125,
	129

C

Cabanis, Pierre Jean Georges 193,
	195, 197
calculus 215
Carlyle, Thomas 166
Carter, K. Codell 137
Catholic Church 169, 207
Cats-eyes Cunningham 288
Chain Home radar system (CH)
	220–222, 227, 229, 230, 232, 233,
	234, 236, 238, 273
Chamberlain, Sir Thomas 26
Charenton Hospice (France) 205
Charge of the Light Brigade 161
Charles II 16, 30
Chevingé, M. 192
childbed fever 86, 135–137, 138
chloroform 64
cholera 63, 66, 68–80
Chowder, Ken 243
Churchill College, Cambridge 234,
	240
Churchill, Sir Winston 166, 220, 224,
	225,
Cicero 188

Civil War (English) 91, 269
Collins, Harry 93, 103
Combe, George 208
Constans, Adolphe 186–187
contagionism 66, 72, 73, 79
controlled clinical trials 114, 116,
	125
Cook, Captain James 119, 120, 121,
	125, 129
Copernicus, Nicolaus 5, 16
Coulmiers, François Simonet 205,
	206
Couthon, Georges 192, 196, 197, 199
Cowper, William 108
Cry and the Covenant, The
	(Thompson) 143
Cuvier, Georges 199

D

d'Orléans, Louis-Philippe (King of the
	French) 199, 207
Damadian, Raymond 216
Darwin, Charles 12, 85, 238, 274
Dawkins, Richard 217
De Revolutionibus (Copernicus) 16
death ray 221
Descartes, René 15, 53, 55, 56, 87,
	126, 238, 268
Desgauliers, John Théophile 100,
	102
Deutsch, Albert 191
Devil, the 20, 173, 174–185
digitalis 124
Discoverie of Witchcraft (Scot) 18, 180
Dix, Dorothea 165
Drasche, Professor Anton 75
Drummer of Tedworth, the 20–34
Dubos, René 245
Durand, Ludovic 193
Dutch East India Company 117, 118

E

East India Company 117, 118
Einstein, Albert 238, 265
Elizabeth I 43
Emmerich, Rudolf 74
epigenesis 40, 47, 54, 55, 58–60
Epstein, S. 258
Erasmus of Rotterdam 180, 181, 182
Esquirol, Jean-Etienne 198, 199,
      200, 202, 204, 205, 207, 209
Eureka moments xi, 3, 73, 85, 87, 259
European Witch-craze, The
      (Trevor-Roper) 172
Evans, Richard 71
experimental method, the 22–34,
      91–92, 100–105, 106–107, 269
experimenter's regress 102–104
experimentum crucis, the 92, 93–95,
      99, 101, 102, 104, 107, 110

F

Faraday, Michael 239
father figure, the 1, 238–240
Fathers of the Christian Doctrine 204
Feldman, William 248, 251, 256, 262
Fermi, Enrico 234
Fleming, Alexander 251
Florey, Howard 251
Fourcroy, Antoine 195, 197
Franco-Prussian War 77
French Revolution, the 167, 192,
      193, 204, 209
Frias, Alonzo Salazar y 180
Fry, Elizabeth 165

G

Galileo, Galilei 16, 23, 85, 161, 238

Gall, Franz Joseph 207, 208
Gallo, Robert 216
Garden of Eden 181
Garden, George 56
Garrison, F. H. 161
Gascoines, John 96, 97
generation 32, 39–52, 55–60
George III 201
germ theory of disease 9, 65, 66, 124,
      152–154, 158
German Unification 79–80, 270
Glanvill, Joseph 9–12, 15–36, 54, 88,
      91, 174, 268–269
   on scientific method 15, 22–28, 32,
      174
   investigates witchcraft 25–28
   Scepsis Scientifica 17, 35
   arguments for existence of witches
      33–36
   A Philosophical Endeavour 16
   Vanity of Dogmatising 15, 35
   Some Philosophical Considerations
      20
   A Blow at Modern Sadducism 20
   Saducimus Triumphatus 20, 21, 26,
      29, 32, 33, 35
Goering, Hermann 229
Goldstein, Jan 206
Golem, The (Collins and Pinch) 103
Gordon of Khartoum (Major-General
      Charles George Gordon) 161
Gordon, Andrew 142
Gorget, Etienne Jean 210
Gortvay, György 137
Gould, Stephen Jay 2, 4
Great Fire of London 118
Great Man history 2, 86
Gresham College, London 15
Greville, Fulke 181
Groves, General Leslie 234
Gustav VI, King of Sweden 251,
      259

H

Habsburgs, the 145–147
Haffekine, Waldemar 72
Haller, Albrecht von 57
Halley, Edmond 215
Hamburg 63, 79
Harvey, William 55
Harvie, David 113
Haslar Naval Hospital 121, 130
Hearst, William Randolph 166
Heaviside layer 227
Hebra, Ferdinand Ritter von 143,
    144, 156
Hecker, Erich 225
Henry, George 170
hero-worship 165–166
Hertz, Heinrich 222, 238, 240
Hinshaw, H. C. 248, 256, 262
History of Medical Psychiatry (Zilboorg
    and Henry) 170
Hitler, Adolf 221
HMS *Centurion* 113
HMS *Endeavour* 119
HMS *Salisbury* 115, 116, 123, 126,
    129, 130
HMS *Suffolk* 130
Holmes, Sherlock 56
Holton, Gerald 32
Homo sapiens 10
Hood, Lord Samuel 129
Hooke, Robert 15, 29, 87, 88, 98–100,
    106, 107, 215, 238
Hopkins Marine Station (California)
    250
Hulme, Nathaniel 118, 130
Hülsmeyer, Christian 223–226, 233,
    273
humoural theory 27, 124–128,
    173–176
Hungarian Revolt 146
Hunt, Robert 18, 26

Hunter, John 45, 64
Huxham, John 118
Huxley, Thomas H. 267
Huygens, Christiaan 54
hysterodemonopathy 186

I

Idea of Progress, The (Bury) 17
Image, The (Boorstin) 165
imaginal disks 50, 51
immaterial forces 29–31, 53–54
immunity 73
immunology 81
Inquisition, the 169, 170, 179
insanity, nature and causes of
    173–174, 200, 208, 209, 211,
    212
insulin 215
ionosphere 227

J

Jenner, Edward 144
Jesuit Order, the 97, 98, 99
Jones, Doris 257, 258
Jones, Sir R. V. 228, 229, 231
Judgement Day 181

K

Keats, John 93
Keeler, Christine 166
Kepler, Johannes 87, 238
Kierkegaard, Søren 1
King, A. V. 225
King, Martin Luther 165
Klein, Johannes 140, 143, 144, 147,
    155, 156, 271

Koch, Robert 63, 64, 66, 136, 154, 159, 161, 271, 275
  reputation 68–69
  Koch's postulates 68–71
  identifies tuberculosis bacillus 68–71
  relations with Pasteur 78
  studies cholera 68–69
  national hero 70, 77–80
  problems with cholera research 71–73
  challenges Pettenkofer 75
Kolletschka, Professor 139, 140
Kossuth, Lajos 146
Kramer, Heinrich 169, 175
Krebs, Hans 251
Krieff, Paul de 63, 64, 65
Künhold, Rudolph 228, 230, 231, 232, 239, 240

L

laboratory culture 216–217, 263
Lamarck, Jean-Baptiste 199
Lancaster, Sir James 117
Lawrence, Christopher 115, 132
Lawrence, Ernest 234
Leibniz, Gottfried 54, 215, 216
letters de cachet 193
Liebig, Justus von 150–151, 154, 155, 158
Liège seminary 96, 99, 104
light, theories of 92, 93–95, 98–100
Lincoln, Abraham 165
Lind, James 3, 86, 113–161, 167, 269, 270
  1747 experiment 86, 114–116, 120, 125–126, 129–132
  reputation 114–115
  career 116–117, 120–121

Treatise on Scurvy 117, 119, 127, 120, 121, 125, 131, 270
  difficulties with experiment 124
  suggested cures 127–132
  on the properties of citrus fruits 125–132
  mechanical philosophy 126–129
Line, Francis 96, 104, 105
Linnaeus, Carolus 42
Lister, Baron Joseph 136, 159
localists 66
Locke, John 29
Lord of the Rings (Tolkein) 165
Loudon, Irvine 137
Louis XVI (King of France) 195
Lucas, Anthony 96, 101, 105, 106, 107
Ludgarshal (England) 20, 21
Luftwaffe, the 215
Lumpe, Eduard 151
lunaria 173
Luther, Martin 174, 181, 182
Lutheranism 181, 182

M

MacBride, David 121, 122, 125
magicians 176–177
Magnetic Resonance Imaging (RMI) 216
Malebranche, Nicolas 39, 40, 49, 50, 55, 56
Malleus Maleficarum (Sprengler and Kramer) 169
Malpighi, Marcello 49, 56
Marconi, Gugliemo 222, 238, 239, 240
Mariotte, Edmé 99, 100, 102, 105, 106, 107, 110
Martin, Benjamin 85
Mascon (France) 28

Massachusetts Institute of Technology (MIT) 231
materialism 59
maternal impressions, doctrine of 52
Maxwell, James Clerk 222, 239, 240
Mayo Clinic, Minnesota 248, 251
McBride, William 115
Medawar, Sir Peter 4
Medical Men and the Witch during the Renaissance (Zilboorg) 169, 172
Mentally Ill in America, The (Deutsch) 191
Merck pharmaceuticals 249, 256, 262, 272
Metchnikoff, Elie 74
Metternich, Clemens von 145
Meyer, Arthur 39
miasmas 66, 137–138, 149–152
Michaelis, Gustav Adolph 142
Millingen, J. G. 211
Mompesson, Mr. 20, 25, 26, 27, 28
Montagnier, Luc 216
Montaigne, Michel de 180
moral treatment 200–202, 205
More, Henry 25
More, Thomas 165
Morzine (France) 185–187
Mother Theresa 165
Müller, Charles 191, 192, 195, 199

N

Napoleon III, Emperor 192
National Physical Laboratory 221
Navy's Sick and Hurt Board 64
Nazis 229, 273
Newsholme, Sir Arthur 64
Newton, Sir Isaac 3, 15, 16, 17, 29, 30, 86, 91–110, 215, 216, 216, 268, 269
experiments on light 91–110

experimentum crucis 92, 93–95, 99, 100–102, 104, 107, 110
Eureka moment 85, 87
forces acting at a distance 30, 33, 34, 53, 58
theory of universal gravitation 30, 33, 34, 53, 54, 87–89
difficulties with *experimentum crucis* 100–107
President of the Royal Society 108–110
Opticks 34, 99, 101, 106
Principia Mathematica 54, 87, 88
Nightingale, Florence 159, 165
Nisbet, William 210
Nobel Prize 216, 227, 244, 252, 259, 261, 264

O

occult forces (see immaterial forces)
Office of Sick and Hurt Seamen 114
Old Testament, The 59
On the Generation of Animals (Harvey) 55
Oppenheimer, J. Robert 234
Opticks (Newton) 34, 99, 101, 106
Ovism (*see* preformation)
Oxford Martyrs, The 161
Oxford University 15

P

Pakistan Dental Journal 252
Palliser, Admiral Sir Hugh 122, 123
Paris Academy of Medicine 191
Pasteur, Louis 65, 68, 70, 76, 77–78, 136, 154, 159, 161
pathogenesis 51
Pauli, Wolfgang 11, 191–212

penicillin 243, 255, 272
Pepys, Samuel 29
Perrault, Claude 57
Perry, William (Captain Cook's
    mate) 125
Pettenkofer, Max von 9, 10, 11,
    63–81, 270, 271, 273, 275
  auto-experiment 63, 65, 74, 81
  suicide 64
  fall from grace 64–65
  germ-soil theory 66–67, 72–74,
    75
  evidence for theory 74–76
  politics 195, 198, 203
*Philosophical Transactions of the
    Royal Society* 92
phrenology 207–208, 212, 273
Pickles, E. L. 232, 237
Pied Piper of Hamelin 177
Pinch, Trevor 93, 103
Pinel, Philippe 167, 191–210, 273
  chain breaking 167, 191, 193, 195,
    199
  on Johann Weyer 185
  fall from grace 195
  use of medicines 200, 210–211
  moral treatment 200–202
  rehabilitation 199
  use of chains 200
  phrenology 207–212
  achievements buried beneath a
    myth 207–212
Pinel, Scipion 198, 199, 202, 204,
    205, 207
pneuma 54
preformation 9, 39–52, 55–60
Prelude, The (Wordsworth) 91
presentism xi, 3, 4–6, 22, 60, 80
preternatural 31
Priestley, Joseph 122, 123
Principia Mathematica (Newton) 54,
    87

Principles of Radar (MIT) 231
Pringle, Sir John 121, 122, 126
*Proceedings of the Royal Society of
    Medicine* 136
Profumo, John 166
*Proposal for Preventing the Scurvy in the
    British Navy* (Hulme) 118
psychiatry, field of 185–188,
    205–212, 271–272
pulsars 216
Pussin, Jean-Baptiste 196, 197, 203,
    212
Pussin, Mme. 196

Q

quarantine 69, 79

R

radar 215, 219–240
Raeder, Admiral Erich 228
Reichstag, the 70
Revolution (1848) 145–148, 271
Rise of Embryology (Mayer) 39
Rizzetti, Giovanni 99, 100, 102, 105,
    106, 107, 110
Rochefoucauld-Liancourt, duc de La
    197
Rockefeller Institute (New York) 245
Roddis, Louis 114
Rokitansky, Carl 143, 147, 148, 156
Roosevelt, Franklin D. 166
Rosetta Stone 48
Rousseau, Jean-Jacques 193
Royal Air Force, the 219, 220, 229
Royal Navy 116
Royal Society, The 15, 16, 17, 22, 23,
    24, 25, 28, 29, 35, 88, 91, 93, 95,
    101, 102, 107, 109, 269

Royer-Collard, Antoine-Athanase 205, 206, 209
Rush, Benjamin 211
Rutgers Institute of Microbiology 264
Rutgers University 244

S

Sabin, Albert B. 251
Sadducism 20
Salpêtrière Hospital 195, 197, 198, 199
Sandwich, fourth Earl of 122
Sarum 21
Satan 31, 107
Scanzoni, Wilhelm 150, 151, 154, 155, 158
Schaffer, Simon 93, 100, 110
Schatz, Albert 217, 244–266, 277
  biography 244–246
  studies actinomycetes 247–250
  discovers streptomycin 247–248
  relations with Waksman 250, 253, 254, 256–264
  sues Waksman 250–251
  career compromised 252
  tells his version of events 252
Schatz versus Waksman 250–251, 253, 259, 265
Schlesinger, Arthur M. 165
Schuh, Franz 144, 147
Science Wars, the 275
Scot, Reginald 18, 179
scurvy 113–133
Selfish Gene, The (Dawkins) 217
Semmelweis, Ignaz 3, 86, 135–162, 167, 271, 273
  reputation 135–137
  introduces handwashing 135, 140–141

insanity and death 136, 141–149, 156–157
  cadaverous particles theory 139–140
  political activism 146–147
  difficulties with theory 149–154
  personality 148–149, 154
  Aetiology, Concept and Prophylaxis of Childbed Fever 141, 153, 154
Semmelweis, Maria 161, 157
Semmelweis: His life and doctrine (Sinclair) 135
Seyfert, Bernhard 143
Shakespeare, William 41, 189
Simon, Sir John 66
Simpson, James Young 64
Sinclair, Sir William 135
Skoda, Joseph 143, 144, 148, 152, 156
Smithsonian, the 243, 263
Snow, C. P. 85
Social Contract, The (Rousseau) 193
Spallanzani, Lazzaro 9, 10, 11, 12, 39–60, 88, 268, 269
  reputation 39–41
  experiments on generation 42–47
  mechanical philosophy 55–58
  sperm (role in procreation) 40, 54
Spermists (*see* preformation)
spontaneous generation 40
Sprat, Thomas 23, 92
Sprengler, Jacob 169, 175
SS Graf Spee 229
St. Jerome 174
St. Prix Ward (Bicêtre Hospital) 196, 197
St. Rochus Hospital 141
Star Trek 165
Star Wars 165
Streptomyces griseus 247, 254, 257, 259, 260
streptomycin 215, 243–266, 272

Strong, Dr. A. 258
Suez Canal 69
sulfa drugs 272
supernatural, the 30
Swammderdam, Jan 46, 48, 49, 50,
	55, 56
Swift, Jonathon 50

T

tacit knowledge 100–101
Tacitus 188
Taylor, Albert Hoyt 227, 228, 230,
	231, 239, 240
Tedworth 20, 26, 32
Telemobiloscope 223, 224, 225,
	226
Tesla, Nikola 222, 238
Thévenot, Melchisedec 48
Thompson, Morton 143
Three Steps to Victory (Watson-Watt)
	219, 220, 225
Time (magazine) 243
Tirpitz, Admiral von 224
Tizard, Sir Henry 221, 235, 237,
	240
Trevor-Roper, Hugh 172
Trinity College, Cambridge 93
tuberculosis 68, 71, 77, 215,
	243–266
Tucker, W. S. 235
Tuke family of York, the 203
Tuke, Samuel 211
Tuke, William 203
Tuve, Merle 227, 232
Typhoid Mary 73

U

University of Pest 141

V

Vandermeersch, Patrick 187
Viennese lying-in hospital 135,
	137–140, 144, 153
Virchow, Rudolf 151, 153, 155–156
vitalism 59
Vitamin C 121, 128, 131

W

Wainright, Milton 253, 258, 259
Wakefield, Edward 211
Waksman, Selman 215, 216, 217,
	243–260, 272
	reputation 243–245
	biography 244
	relationship with Schatz 247
	Nobel Prize 251–252
	taken to court 252, 259, 261, 264
	motivations 264–266
	invents sick chicken story 257–259
	relationship with students
		253–259
	My Life with Microbes 261
	The Conquest of Tuberculosis 261,
		262
Wallace, Barnes 232
Wallis, Captain Samuel 119, 120,
	121, 125, 129
Wallgren, Professor A. 260
Warhol, Andy 165
Waterloo, Battle of 222
Watson-Watt, Sir Robert 215,
	216–240, 273
	reputation 219–220
	Three Steps to Victory 219–220, 225,
		228, 230
	achievements 230
	claims status of inventor of radar
		230–234

defines radar 230–231, 233
Royal Commission 232–234, 237
sidelines other claimants 230–240
Watt, James 85
Wellington, Duke of (Arthur
    Wellesley) 220, 256
Weyer, Johann 167, 169–189, 215
    De Praestigiis Daemonum 170,
        174–175, 177, 176, 179, 185, 188
    father of psychiatry 170–172
    ideas about nature of possession
        172–177
    magicians 176–177
    Satan 174–177, 178
    treating possession 175
    influences 18–182
    reinvented by nineteenth-century
        psychiatrists 187–188
Whewell, William 12
Whiston, William 34, 109
Whitman, Charles 39
Whittle, Frank 232
Wilde, Oscar 195
Wilhelm I, Kaiser 70, 77, 271, 275
Wilkins, Arnold, F. 221, 232, 234,
    235–237, 240
Wilkins, John 22

William III, Duke of Berg, Julich and
    Cleves 170
Willis, Francis 210, 203
Windows (to interfere with radar
    systems) 232
witch-craze, the 16, 167–189
witches 16–36, 54, 169–189
Withering, William 124
Wolpert, Lewis 57
Woods Hole Biology Lectures
    (Whitman) 39
Wordsworth, William 85, 91

Y

Yorkshire Retreat, the 203
Youghall, witch of 21
Young, Leo C. 227, 228, 230, 231,
    239, 240

Z

Zilboorg, Gregory 169, 170, 171, 172,
    174
Zoltán, Imre 137